Numerical Techniques for Boundary Element Methods

Edited by Wolfgang Hackbusch

Notes on Numerical Fluid Mechanics (NNFM) Volume 33

Volume 4 Shear Flow in Surface-Oriented Coordinates (E. H. Hirschel / W. Kordulla)

Volume 8 Vectorization of Computer Programs with Applications of Computational Fluid Dynamics (W. Gentzsch)

Volume 9 Analysis of Laminar Flow over a Backward Facing Step (K. Morgan / J. Periaux / F. Thomasset, Eds.)

Volume 11 Advances in Multi-Grid Methods (D. Braess / W. Hackbusch / U. Trottenberg, Eds.)

Volume 12 The Efficient Use of Vector Computers with Emphasis on Computational Fluid Dynamics (W. Schönauer / W. Gentzsch, Eds.)

Volume 13 Proceedings of the Sixth GAMM-Conference on Numerical Methods in Fluid Mechanics (D. Rues / W. Kordulla, Eds.)

Volume 14 Finite Approximations in Fluid Mechanics (E. H. Hirschel, Ed.)

Volume 15 Direct and Large Eddy Simulation of Turbulence (U. Schumann / R. Friedrich, Eds.)

Volume 16 Numerical Techniques in Continuum Mechanics (W. Hackbusch / K. Witsch, Eds.)

Volume 17 Research in Numerical Fluid Dynamics (P. Wesseling, Ed.)

Volume 18 Numerical Simulation of Compressible Navier-Stokes Flows (M. O. Bristeau / R. Glowinski / J. Periaux / H. Viviand, Eds.)

Volume 19 Three-Dimensional Turbulent Boundary Layers - Calculations and Experiments (B. van den Berg / D. A. Humphreys / E. Krause / J. P. F. Lindhout)

Volume 20 Proceedings of the Seventh GAMM-Conference on Numerical Methods in Fluid Mechanics (M. Deville, Ed.)

Volume 21 Panel Methods in Fluid Mechanics with Emphasis on Aerodynamics (J. Ballmann / R. Eppler / W. Hackbusch, Eds.)

Volume 22 Numerical Simulation of the Transonic DFVLR-F5 Wing Experiment (W. Kordulla, Ed.)

Volume 23 Robust Multi-Grid Methods (W. Hackbusch, Ed.)

Volume 25 Finite Approximation in Fluid Mechanics II (E. H. Hirschel, Ed.)

Volume 26 Numerical Solution of Compressible Euler Flows (A. Dervieux / B. van Leer / J. Periaux / A. Rizzi, Eds.)

Volume 27 Numerical Simulation of Oscillatory Convection in Low-Pr Fluids (B. Roux, Ed.)

Volume 28 Vortical Solutions of the Conical Euler Equations (K. G. Powell)

Volume 29 Proceedings of the Eighth GAMM-Conference on Numerical Methods in Fluid Mechanics (P. Wesseling, Ed.)

Volume 30 Numerical Treatment of the Navier-Stokes Equations (W. Hackbusch / R. Rannacher, Eds.)

Volume 31 Parallel Algorithms for Partial Differential Equations (W. Hackbusch, Ed.)

Volume 32 Adaptive Finite Element Solution Algorithm for the Euler Equations (R. A. Shapiro)

Volume 33 Numerical Techniques for Boundary Element Methods (W. Hackbusch, Ed.)

Volumes 1 to 3, 5 to 7, 10, and 24 are out of print.

Numerical Techniques for Boundary Element Methods

Proceedings of the Seventh GAMM-Seminar,
Kiel, January 25-27, 1991

Edited by
Wolfgang Hackbusch

Die Deutsche Bibliothek – CIP-Einheitsaufnahme

Numerical techniques for boundary element methods / ed. by Wolfgang Hackbusch.

(Notes on numerical fluid mechanics; Vol. 33)
ISBN 978-3-528-07633-7 ISBN 978-3-663-14005-4 (eBook)
DOI 10.1007/978-3-663-14005-4

NE: Hackbusch, Wolfang [Hrsg.]; GT

Originally published by Friedr. Vieweg & Sohn Verlagsgesellschaft mbH, Braunschweig / Wiesbaden in 1992
Softcover reprint of the hardcover 1st edition 1992

Produced by W. Langelüddecke, Braunschweig
Printed on acid-free paper

ISSN 0179-9614
ISBN 978-3-528-07633-7

F o r e w o r d

The GAMM Committee for "Efficient Numerical Methods for Partial Differential Equations" organises workshops on subjects concerning the algorithmic treatment of partial differential equations. The topics are discretisation methods like the finite element and the boundary element method for various types of applications in structural and fluid mechanics. Particular attention is devoted to the advanced solution methods.

The series of such workshops was continued in 1991, January 25-27, with the 7th Kiel-Seminar on the special topic

"Numerical techniques for boundary element methods"

at the Christian-Albrechts-University of Kiel. The seminar was attended by 57 scientists from 8 countries.

The list of topics contained applications of the boundary element method (BEM) to various problems of practical interest, algorithmic aspects of the BEM (coupling with finite element method, parallelisation), convergence analysis, and in particular the treatment of the numerical integration. In six contributions the quadrature of weakly singular, Cauchy singular, and hypersingular integrals is analysed.

The editor thanks the "DFG-Schwerpunkt Randelementmethoden" for its support. He also likes to express his gratitude to all persons involved in the organisation of the seminar.

Kiel, June 1991 W. Hackbusch

CONTENTS

On Parallel Processing in 3-D Acoustic BEM

H. Antes
TU - Braunschweig
Institute for Applied Mechanics
Abt - Jerusalemstr. 7
D - 3300 Braunschweig Germany

K. Volk
IBM Scientific Center
Institute for Supercomputing and Applied Mathematics
Tiergartenstr. 15
D - 6900 Heidelberg, Germany

Summary

The propagation of acoustic waves is governed by the scalar wave equation. In this paper, the initial boundary value problem with homogenous Neumann boundary conditions is investigated. Kirchhoff's integral equation yields a time - dependent boundary integro - differential equation for the unknown Cauchy data, which is solved by a boundary element Collocation method. The resulting linear system of equations is given by a lower triangular block Toeplitz matrix, where each block matrix corresponds to a certain time step.

Building up these blocks, which depend recursively with recursion depth 1 on each other, is the most time consuming part of the procedure. For improving this, a parallel algorithm on a shared memory system for the assembly of the block matrices with respect to the time steps is developed. Numerical examples demonstrate a good performance of the procedure.

0. Introduction

During the last decade the boundary element method (BEM) for solving strongly elliptic partial differential equations has become more and more common.

Here, a boundary integral equation method for the hyperbolic time dependent scalar wave equation is presented. In contrast to elliptic problems there are many open questions to the mathematical verification and the efficiency of the boundary element method. In the special case of 3 - D wave equation the mathematical analysis for the Galerkin method has been performed by Ha Duong [4], but here a collocation method, where the complete analysis is still open, is considered. However, numerical results achieved by Th. Meise [1; 2; 9] show high accuracy of the underlying procedure.

Let Ω be a Lipschitz domain in $\mathbb{R}^3$. We look for a solution p of the initial boundary value problem in the space time cylinder $\Omega_T := \Omega x[0, T]$:

$$\begin{aligned} -\frac{1}{c^2} p_{tt}(x,t) + \Delta_x p(x,t) &= S(x,t) \\ p(x,0) &= 0 \text{ for all } x \in \Omega \\ p_t(x,0) &= 0 \text{ for all } x \in \Omega \\ \partial_\nu p\,(x,t) &= 0 \text{ for all } x \in \Gamma := \partial\Omega \text{ and for allmost all } t \in [0,T], \end{aligned} \tag{0.1}$$

where ∂_ν denotes the derivation with respect to the outer normal on Γ, and the subscript t the time derivative. $S(x,t)$ is a given point source and c the speed of sound. A fundamental solution $F(y, \tau; x, t)$ satisfying

$$-\frac{1}{c^2} F_{\tau\tau}(y, \tau; x,t) + \Delta F\,(y, \tau; x,t) = -\delta(x-y)\delta\,(t-\tau)$$

for all times t, $\tau \in \mathbb{R}$ and all x, $y \in \mathbb{R}^3$, where δ is the Dirac functional, is given by [5] :

$$F(y, \tau; x,t) := \frac{1}{4\pi\,|x-y|}\,\delta(t_R - \tau),$$

where the retarded time t_R is defined by $t_R := t - |x-y|/c$. Assuming zero internal source density, except point sources S, and zero initial values, this fundamental solution yields Kirchhof's integral equation for the unknown pressure $p(x,t)$ on the boundary :

$$\begin{aligned} 4\Pi\, d(x)\, p(x, t) &= \int_\Gamma \partial_\nu p(y, t_r) \frac{1}{r}\, d\Gamma_y\; + \\ &\int_\Gamma \left(p(y, t_r) \frac{1}{r^2} + \dot{p}(y, t_r) \frac{1}{cr} \right) \frac{\nu \bullet r}{r}\, d\Gamma_y\; - \\ &\sum_k S_k(t - |x_k - x|/c) \frac{1}{|x_k - x|}\,. \end{aligned} \tag{0.2}$$

This equation holds for all $x \in \mathbb{R}^3$ with

$$d(x) := \begin{cases} 1 & ; x \in \Omega \\ d^i & ; x \in \Gamma \\ 0 & ; x \in \Omega^C, \end{cases}$$

where d^i is the degree of the space angle, which is cut off by the interior of Ω, see [2; 6; 9]. In case of smooth boundary parts d^i is equal 1 / 2. The pressure and the flux on the boundary are represented again by $p(x,t)$ and $\partial_\nu p(x,t)$. The distance between the observation point x and the integration point y is denoted by $r := |x - y|$. Please notice, that the pressure $p(y,t_R)$ and the flux on the boundary $\partial_\nu p(y,t_R)$ are to be set zero for retarded times $t_R < 0$.
Due to (0.2), the solution of the above defined initial boundary value problem is completely characterized by the values of the known and unknown pressure $p(x,t)$ on the boundary $\Gamma x[0,T]$.

1. A time - stepping boundary element collocation method

As described in [1; 2; 9], for the numerical approximation of Kirchhoff's integral equation (0.2) a boundary element collocation method is used. For simplicity assume that Ω is the union of Q disjoint plane surface pieces, $\Omega = \bigcup_{q=1}^{Q} \Gamma_q$. The space discretisation is performed by using a quasi - uniform family of triangulations $\mathcal{I}_E := \{T_e \mid e = 1, ..., E\}$ with

$$\Gamma = \dot{\bigcup_{e=1}^{E}} T_e.$$

The time interval $[0,T]$ is subdivided into M equidistant sub - intervals of length $\Delta t := T/M$, where the product $c\,\Delta t$ is assumed to be larger than or equal to the maximum diameter h_{max} of the circumcircles of the triangles T_e divided by 2. Furthermore, the unknown function $p(x,t)$ is assumed to be restricted on the finite dimensional tensor product space $\mathcal{H}_{NM} := \mathbf{S}_N(\mathcal{I}_E) \otimes S_M(\Delta t)$, where $\mathbf{S}_N(\mathcal{I}_E)$ is the space spanned by the usual linear Lagrange finite element functions on the space grid $\mathcal{I}_E$, and $S_M(\Delta)$ is the space given by the smoothest splines of degree 1 on the time grid $\{m \times \Delta t \mid m = 1, ..., M\}$. This yields

$$p_{NM}(x,t) := \sum_{j=1}^{N} p_j(t)\, \phi_j(x), \tag{1.1}$$

where $p_j(t) \in S_M(\Delta t)$ is a linear combination of B - splines, and $\phi_j(x)$ is a finite element basis function of $\mathbf{S}_N(\mathcal{I}_E)$. Notice, that the polynomial $p_j(t)$ is unique determined by it values at the time knots $m\Delta t$. For $(m-1)\Delta t \leq t \leq m\Delta t$ the function $p_j(t)$ is given by linear interpolation

$$p_j(t) = \left[1 - \frac{(t_m - t)}{\Delta t}\right] p_j(t_m) + \frac{(t_m - t)}{\Delta t} p_j(t_{m-1}).$$

Now, an approximation of (0.2) is be obtained by using $p_{NM}(x,t)$ instead of $p(x,t)$ and by collocation at the knots of the space grid as well as of the time grid:

$$\begin{aligned} 4\Pi\, d(x_i)\, p_{NM}(x_i, t_m) = & \int_\Gamma \partial_\nu p_{NM}(y, t_{R_m}) \frac{1}{r_i} d\Gamma_y + \\ & \int_\Gamma \left(p_{NM}(y, t_{R_m}) \frac{1}{r_i^2} + \dot{p}_{NM}(y, t_{R_m}) \frac{1}{cr} \right) \frac{\nu \bullet r}{r} d\Gamma_y - \\ & \sum_k S_k(t_m - |x_k - x_i|/c) \frac{1}{|x_k - x_i|} \qquad i = 1, ..., N \;;\; m = 1, ..., M. \end{aligned} \tag{1.2}$$

Obviously, an additional space dependence of the integrals results from the evaluation at the retarded times $t_{R_m} = t_m - r/c$. But, because of the assumption $p_{NM}(x,t) = 0$ for $t \le 0$, the integration must performed only for those boundary points $y \in \Gamma$, which fulfill $t_{R_m} = t_m - r_i/c > 0$ for the certain time step t_m. Vice versa, if we consider collocation points, which have a smaller distance as a given fixed value r_0 to integration points y, the influencing time steps t_m fulfill $t_m \ge r_0/c$.

Applying these facts, the linear system of equations (1.2) can be rewritten by elementary computations in the form [1; 2; 9]

$$4\Pi \begin{vmatrix} \vec{\mathbf{d}} \times \vec{\mathbf{p}}(t_1) \\ \vec{\mathbf{d}} \times \vec{\mathbf{p}}(t_2) \\ \vec{\mathbf{d}} \times \vec{\mathbf{p}}(t_3) \\ \cdot \\ \cdot \\ \cdot \\ \vec{\mathbf{d}} \times \vec{\mathbf{p}}(t_M) \end{vmatrix} = \begin{vmatrix} \mathbf{H}_1 & 0 & 0 & \cdot & \cdot & \cdot & 0 \\ \mathbf{H}_2 & \mathbf{H}_1 & 0 & \cdot & \cdot & \cdot & 0 \\ \mathbf{H}_3 & \mathbf{H}_2 & \mathbf{H}_1 & \cdot & \cdot & \cdot & 0 \\ \cdot & \cdot & \cdot & \cdot & \cdot & \cdot & \cdot \\ \cdot & \cdot & \cdot & \cdot & \cdot & \cdot & \cdot \\ \cdot & \cdot & \cdot & \cdot & \cdot & \cdot & \cdot \\ \mathbf{H}_M & \mathbf{H}_{M-1} & \mathbf{H}_{M-2} & \cdot & \cdot & \cdot & \mathbf{H}_1 \end{vmatrix} \begin{vmatrix} \vec{\mathbf{p}}(t_1) \\ \vec{\mathbf{p}}(t_2) \\ \vec{\mathbf{p}}(t_3) \\ \cdot \\ \cdot \\ \cdot \\ \vec{\mathbf{p}}(t_M) \end{vmatrix} - \begin{vmatrix} \vec{\mathbf{f}}(t_1) \\ \vec{\mathbf{f}}(t_2) \\ \vec{\mathbf{f}}(t_3) \\ \cdot \\ \cdot \\ \cdot \\ \vec{\mathbf{f}}(t_M) \end{vmatrix}, \tag{1.3}$$

where the unknown vector $\vec{\mathbf{p}}(t_m)$ at the time step m is given by the values of the coefficient polynomials $p_j(t)$ in (1.1) at the time knots t_m,

$$\vec{\mathbf{p}}(t_m) := (p_1(t_m), ..., p_N(t_m))^T.$$

The vector $\vec{\mathbf{d}}$ is defined by $\vec{\mathbf{d}} := (d(x_1), ..., d(x_N))^T$, where $\times$ denotes pointwise multiplication of the components. The right hand side is related to $\vec{\mathbf{f}} := \left(\sum_k S_k(t_m - |x_k - x_1|/c)/|x_k - x_1|, ..., \sum_k S_k(t_m - |x_k - x_N|/c)/|x_k - x_N|\right)^T,$

and the $N \times N$ block matrices $\mathbf{H}_m$ are given by

$$(\mathbf{H}_m)_{i,j=1,\dots,N} := \left(m \int_{\Gamma^m} \frac{\nu \bullet r_i}{r_i^2} \, \phi_j(y) \, d\Gamma_y + (2-m) \int_{\Gamma^{m-1}} \frac{\nu \bullet r_i}{r_i^2} \, \phi_j(y) \, d\Gamma_y \right)_{i,j=1,\dots,N} \tag{1.4}$$

$m = 1, \dots, M,$

with

$$\Gamma^m := \{ y \in \Gamma \mid c(m-1)\Delta t < |x - y| \le cm\Delta t \}$$

and $r_i := |x_i - y|$. As to be seen from (1.4), the block matrices $\mathbf{H}_m$ may have many zero entries. The portion of these zeros depends on the ratio $(c\Delta t)/h_{max}$. For example, Figure 1 shows the matrix $\mathbf{H}_1$ for 872 finite elements in the space variables and $c\Delta t = h_{max}$, where the domain Ω is a cuboid. The non - zero entries are printed as black points.

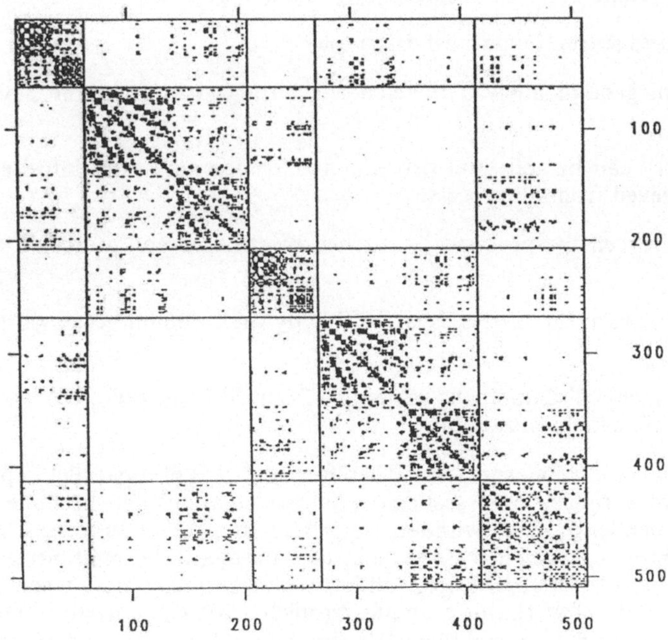

Figure 1. : Non zero matrix entries

Now a parallel algorithm for the assembly of the whole stiffness matrix (1.3) is formulated.

2. The parallel algorithm

In the beginning of this section, a brief description of key definitions [8; 10] is given, when a *shared memory* machine for the implementation of the algorithm is used. A shared memory machine has a single global memory accessible to all processors. Each processor may have in addition a local memory such as the cache on the IBM ES/3090 or IBM ES/9000. A data in the global memory can be fetched from each processor. Parallelism on a shared memory architecture is to be understand as the execution of several *processes, i.e. of independently executable code sequences,* on different processors at the same time. A programming model is given by:

- A initializing process starts the parallel execution.
- Each process is in general allowed to fetch a arbitrary data from the global memory.
- The global memory can be split into private data for a process and into data, which can be achieved from all processes.
- The connection between the processes is reached by the common access on the global memory.
- Synchronization between the processes is possible by the common access on the global memory.
- The access to the identical data at the same time from different processes has to be resolved serial via synchronization.

Synchronization may be relatively expensive. Therefore, the goal of a parallel implementation is to split of a process into sections, where calculations can be done as much as possible independently, i.e. without using data from other processes. An algorithm has a *fine grain approach* if these sections contain relatively small pieces of work, for example parallel execution of loops, and has *coarse grain approach* when the sections contain large pieces of work, for example, parallelization on subroutine level. There are several general guidelines including the above stated, from a performance standpoint, for designing parallel programs. Some of these are [7] :

1. Minimize the work, that must done serially.
2. Minimize the overhead due to executing parallel constructs.
3. Minimize storage contention. This introduces dependencies on both the architecture and implementation of the program running system. In general terms, this item can be stated as:

- minimize access to storage by taking maximum use of data already in local CPU storage and
- avoid inter - CPU conflicts in the system storage access mechanisms.

4. Assign balanced work loads.

Now the calculation of (1.4) will be analyzed with respect to this four points. In order to satisfy the first point let us consider the theoretical complexity of our procedure. Assume the numbers of operations needed for the the evaluation of one integral in (1.4) for a fixed finite element basis function and a fixed collocation point is of order $O(1)$. Then, the assembly of the stiffness matrix has complexity $O(N^2)$.
For solving the arising linear system of equations $O(BMN) + O(B^2N)$ operations are required, when we take the LU - decomposition of $\mathbf{H}_1$ and resolve then the triangular block system by forward substitution. The integer B is the band width after rearranging the matrices $\mathbf{H}_m$ as band matrices. Because the band width is independent of N for $c\,\Delta t\,/\,h_{\max} = const.$ the assembly of the stiffness matrix is the most time consuming part of the algorithm. Figure 5 in Appendix A confirms this observation in terms of a real application. Therefore, the main task is to find a parallel approach for computing the matrices $\mathbf{H}_m$. For solving the system of equations in a way described above, scientific program library subroutines using a parallel implementation of BLAS 3 routines [3] are be proposed.
Investigating the data dependencies which occur by the definition of the block matrices, we see from (1.4), that each $\mathbf{H}_m$ is completely independent from the others. Moreover, the result of the computation of the integral over Γ_m for the matrix $\mathbf{H}_m$ can be used for the assembly of $\mathbf{H}_{m+1}$, because the first integral in definition (1.4) of $\mathbf{H}_m$ and the second integral appearing in definition (1.4) of $\mathbf{H}_{m+1}$ are equal. This procedure reads as follows:

(a) Initializing : Let $\mathbf{H}_m = 0$ for all $m = 1, \ldots, M$.
(b) Initializing : Let $m = 1$.
(c) Compute the matrix $\mathbf{INTm} = \left(\int_{\Gamma^m} \frac{\nu \bullet r_i}{r_i^2} \, \phi_j(y) \, d\Gamma_y \right)_{i,j=1, \ldots, N}$
(d) Let $\mathbf{H}_m = \mathbf{H}_m + m \times \mathbf{INTm}$
(e) Let $\mathbf{H}_{m+1} = \mathbf{H}_{m+1} + (1 - m) \times \mathbf{INTm}$
(f) Let $m = m + 1$
(g) If $m \leq M$ continue with (c).

Notice, that the computation of $\mathbf{H}_m$ in the steps (c) - (g) for a fixed m depends on the steps (c) - (g) for $m - 1$. This recursion dependence has recursion depth 1. The data dependence between two subsequent step series (c) - (g) is the result of this action, which must be considered for the synchronization of the parallel program. As a result of this required synchronization patter, an additional serial work may occur.
For the parallel assembly of the stiffness matrix, a coarse grain approach is used, where several block matrices $\mathbf{H}_m$ are be built up in each parallel task. Because the block matrices are stored subsequently with increasing m, this approach is also very suitable in the underlying shared memory architecture to obtain a good use of local CPU storage and to avoid inter - CPU storage conflicts.

In order to assign balanced work loads, the execution time of the loop (c) - (g) for different m has to be estimated. In case of large geometries in space direction the work for all m is nearly the same, because the intersection of all Γ_m with Ω can here be assumed having the same area. This is not true for inhomogenous domains Ω. The execution time for different m may be vary in a large range.

Taking the conclusion of the discussion of the general features of our algorithm, one can state that the points 1. 2. and 4. influence each other in the following way :
In order to get minimal overhead of the parallel constructs it is required to handle as much possible different block matrices in each parallel task. But, this is for inhomogenous domains Ω in contradiction with good balanced work loads. Further, the work for the synchronization will increase.
Therefore, two different algorithms are proposed. The first one is designed for large geometries, where the work load problems do not occur. The second one take regard to the unbalanced work loads of the steps (c) - (g), occurring in general. Let P the number of processors, and let M the number of time steps to be computed. For simplicity M/P is assumed to be an integer.

ALGORITHM 1:

1. Divide the series of matrices $\mathbf{H}_1$, ..., $\mathbf{H}_M$ into P parts of length M/P and schedule the work as following over P processes

   - PROCESS 1 : $\leftarrow$ $\mathbf{H}_1, \ldots, \mathbf{H}_{M/P}$

     .

     .

     .

   - PROCESS P : $\leftarrow$ $\mathbf{H}_{M/P(P-1)+1}, \ldots, \mathbf{H}_M$,

   i.e. PROCESS i, $i \in \{1, \ldots P\}$, handle the code sequences (c) - (e) mentioned above for $m \in \{(i-1)M/P+1, \ldots, iM/P\}$.

2. To resolve a possible access on the data $\mathbf{H}_{(i-1)\times M/P+1}$ of PROCESS (i-1) and PROCESS i at the same time synchronization between the two subsequent processes is required. PROCESS (i-1) is forced to wait until PROCESS i has finished the execution of the steps (c) - (e) for $m = (i-1)M/P + 1$. This is necessary for all PROCESSes i, with $i = 1, \ldots, P-1$.

3. The program continues with solving the algebraic system, when all processes have finished their work.

ALGORITHM 2:

1. Divide the series of matrices $\mathbf{H}_1$, ..., $\mathbf{H}_M$ into parts of length 4 and a rest with length smaller than 4. Schedule the work as following over $[M/4]$ processes if $M/4$ is an integer, respectively $[M/4]+1$ processes if $M/4$ is not an integer

   - PROCESS 1 $: \leftarrow$ $\mathbf{H}_1, ..., \mathbf{H}_4$

     .

     .

     .

   - PROCESS [M/4] $: \leftarrow$ $\mathbf{H}_{4([M/4]-1)+1}, ..., \mathbf{H}_{4[M/4]}$
   - PROCESS [M/4] + 1 $: \leftarrow$ $\mathbf{H}_{4[M/4]+1}, ..., \mathbf{H}_M$,

   i.e. PROCESS i, $i \in \{1, ..., [M/4]\}$, handle the code sequences (c) - (e) mentioned above for $m \in \{4 \times (i-1)+1, ..., 4 \times i\}$ and PROCESS [M/4] + 1 handle (c) - (e) for $m \in \{4[M/4]+1, ..., M\}$.

2. To resolve a possible access at the same time on data $\mathbf{H}_{(i-1)\times 4+1}$ of PROCESS (i-1) and PROCESS i synchronization between the two subsequent processes is required again. PROCESS (i-1) is forced to wait until PROCESS i has finished the execution of the steps (c) - (e) for $m = 4 \times (i-1)+1$. This is necessary for all PROCESSes i, with $i = 1, ..., [M/4]$.

3. The program continues with solving the algebraic system, when all processes have finished their work.

Please notice, that the probability of data access at the same time for two subsequent processes in ALGORITHM 1 decreases with increasing number of time steps M.

3. Numerical experiments

Let the domain Ω be a noise protection wall of 50 meter length and 4 meter height. A triangulation with 432 triangles is issued, where the ratio $(c\Delta t)/h_{max}$ is chosen equal to 0.51. $M = 25$ time steps are computed.
ALGORITHM 1 has run on a dedicated IBM 3090 - 600 J under VM / XA at TU - Brauschweig for the above given geometry. Figure 2 shows the required execution time for the assembly of the whole stiffness matrix in dependence on the number of used real CPU's.

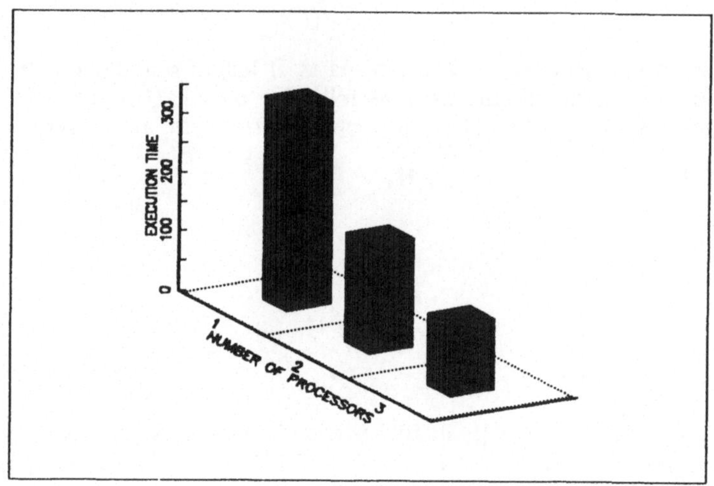

Figure 2. : Execution time

The speed - up := $\frac{\text{scalar execution time}}{\text{parallel execution time}}$ for one, two and three real CPU's is shown in Figure 3. The speed - up is nearby the optimal value which can be achieved in any case.

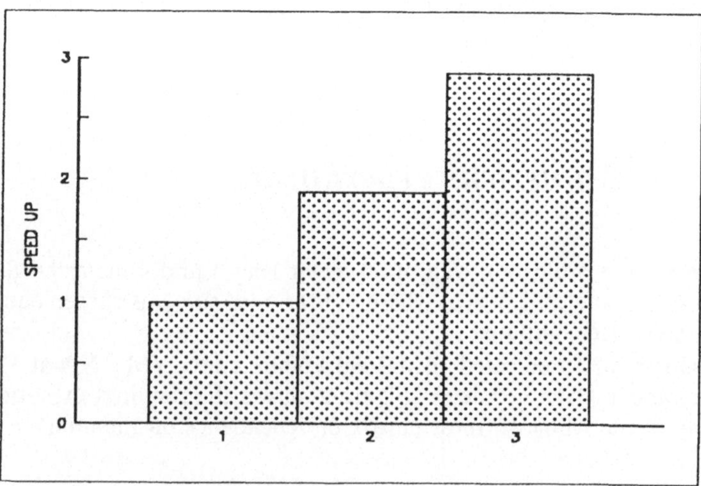

Figure 3. : Parallel speed - up

ALGORITHM 1 has also run on an IBM 3090 - 400 E under MVS/ ESA in concurrence with other users. Here the performance of ALGORITHM 1 is not so good as on the dedicated system, because unbalanced work load may occur, when a required real CPU for a process is working for another user and can therefore not be used. The so delayed process is forced to do work in serial, because in ALGORITHM 1 the number of used processes and real CPU's are the same. Figure 4 describes this situation in terms of parallel speed - up's.

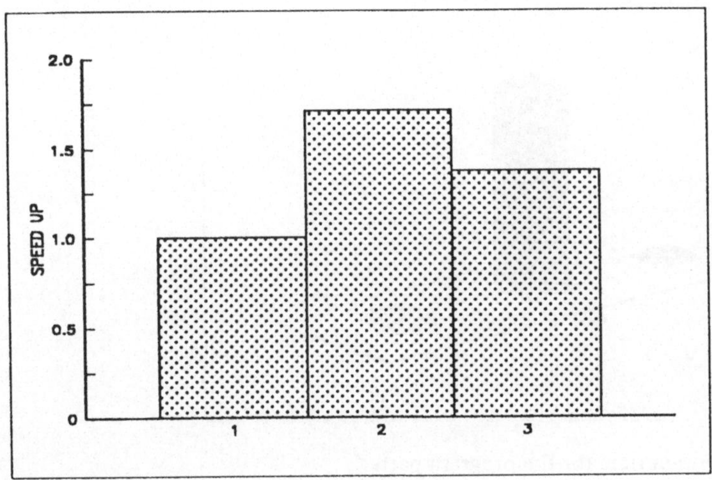

Figure 4. : Overloaded 3090 - 400 E speed up

In contrast to the first version, ALGORITHM 2 is able to spread more work on the other CPU's if processes are delayed on a certain CPU. Therefore we got here a speed - up factor 2.4 on the same overloaded IBM 3090 - 400 E. This shows, that ALGORITHM 2 is superior to the first algorithm on a non - dedicated system. Unfortunately, we are not able to present run time examples on a dedicated system for ALGORITHM 2, but the results on the IBM 3090 - 400 E together with other users are much better as for ALGORITHM 1 also for inhomegenous regions Ω. This seems very promising for the future.

Appendix A

Time needed for initializing, matrix assembly and solving the system of equations is shown in parts of percent of execution time. The number of time steps considered is $M = 20$, and the domain Ω was discretized by 128 triangles.

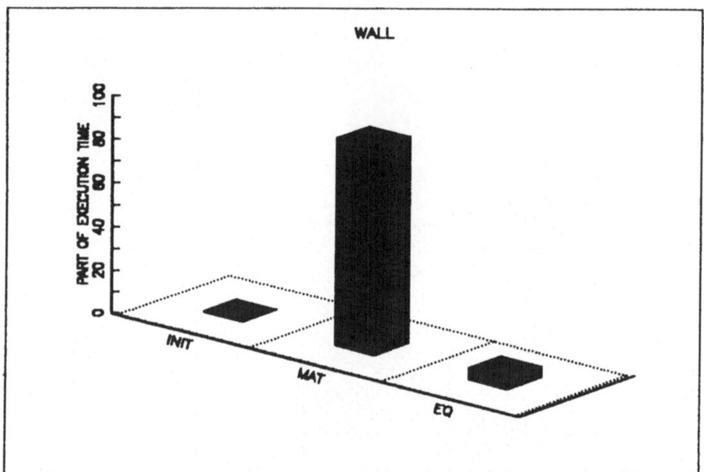

Figure 5. : Execution time for the program parts

Acknowledgements

The tool to generate the representation of the matrix was kindly provided by G. Paolini & P. Santangelo from IBM ECSEC, Rome

We thank the RZ team of TU - Braunschweig for kindly advising and support of the numerical tests on the IBM 3090 - 600 J. The whole work is part of the EASI.

References

[1] **H. Antes** : *Application in Environmental Noise.* Chapt. 11 in : Advances in BEM in Acoustics, Eds. : R. Ciskowski, C. A. Brebbia, Comp. Mech. Publications, Southampton (1991).

[2] **H. Antes, Th. Meise :** *Scalar wave propagation - calculation capabilities of a 3D time domain boundary element method.* In: Boundary Elements XI , Vol. 4 (C. A. Brebbia ed.), Springer - Verlag Berlin (1989).

[3] **C. Bischof, J. Demmel, J. Dongarra, J. Du Croz, A. Greenbaum, S. Hammarling, D. Sorensen :** *LAPACK Working Note #5 : Provisional Contents.* Argonne National Labratory, ANL - 88 - 38 (1988).

[4] **M. T. Ha Duong :** *Equations integrales pour resolution numerique de problemes de diffraction d'ondes acoustiques dans* $\mathbb{R}^3$. These de doctorat d'etat es sciences mathematiques, Univ. Pierre et Marie Curie Paris (1987).

[5] **P. H. L. Groenenboom :** *The Application of BE of Steady and Unsteady Potential Fluid Flow Problems in 2D and 3D.* In : Proc. BEM 3 (Ed. Brebbia), Springer, Berlin (1981).

[6] **F. Hartmann :** *The Somigliana identity on piecewise smooth surfaces.* Journ. of Elast., Vol. 11, 403 - 423 (1981).

[7] *IBM Parallel Fortran, Language and Library Reference (1988).*

[8] **A. H. Karp :** *Programming for Parallelism.* IEEE Computer Society Press, 43 - 57 (1987).

[9] **T. Meise :** *Randelementverfahren zur Berechnung der Ausbreitung skalarer Wellen im 3D - Zeit- und Frequenzbereich.* PhD thesis, Univ. Bochum (1990).

[10] **W. Rönsch, R. Reuter :** *Lineare Algebra für Parallelrechner.* To appear in : Praxis der Informationsverarbeitung und Kommunikation, Vol. 3 (1991).

ON THE BOUNDARY ELEMENT METHOD REALIZATION ON A TRANSPUTER SYSTEM

by K.GEORGIEV

Center for Informatics and Computer Technology

Acad.G.Bonchev str., Bl.25 A, 1113 Sofia, Bulgaria

SUMMARY

In this paper a parallel algorithm for the direct boundary element method (BEM) and its realization on a transputer based computer system are presented. This is connected with two problems in the fluid mechanics. The first one is about an investigation of a potential flow around a profile and second one - about an ideal flow in channels with an arbitrary shape of the walls. A piecewise linearly approximation of the boundary of the domain and constant elements are used. A parallel version of Gaussian elimination method is used to solve the system of linear algebraical equations (SLAE). Some results about the execution time and speed-up for a model problem are presented.

INTRODUCTION

Many problems in physical sciences and engineering can be formulated as boundary value problems of partial differential equations. A large class of elliptic boundary value problems can be presented as integral equations only on the boundary of the domain. This is the class of problems when the fundamental solution of the differential equations can be found explicitly. In recent years boundary element method (BEM) have been developed by many authors (Hsiao, Wendland, Brebia, Butterfield, Nedelec, etc.). There are several important advantages of BEM with respect to the most popular finite element method (FEM). One of these advantages, maybe the most important one, is that BEM requires only a discretization on the boundary rather the domain. This method is suitable especially for external boundary value problems in infinite domains. Unfortunately, the coefficient matrices appeared in numerical treatments are fully distributed and often nonsymetric.

In the last years the transputer based parallel machines are becoming very popular because of their high potential performance. Multi-Transputer systems are being used increasingly in a wide variety of application areas. Most of these use work stations or personal computer as a host system which provides file server and network services. Transputer based computers are networks of processors without common memory, which communicate by dedicated links [BK].

In this paper we present a parallel algorithm of direct BEM and its realization on the distributed computer system including four transputers T 800.

B E M FOR THE LAPLACE EQUATION IN $\mathbb{R}^2$.

Let $\Omega \subset \mathbb{R}^2$ is a bounded domain in the plane with a smooth boundary $\Gamma = \Gamma_1 \cup \Gamma_2$. We will consider the next model problem:

$$\Delta u = 0, \quad (x,y) \in \Omega, \tag{1}$$

$$u|_{\Gamma_1} = u_1(s), \tag{2a}$$

$$q|_{\Gamma_2} = \frac{\partial u}{\partial n}\Big|_{\Gamma_2} = q_2(s). \tag{2b}$$

where $\frac{\partial}{\partial n}$ is the exterior normal derivative on the boundary Γ of Ω (Fig.1).

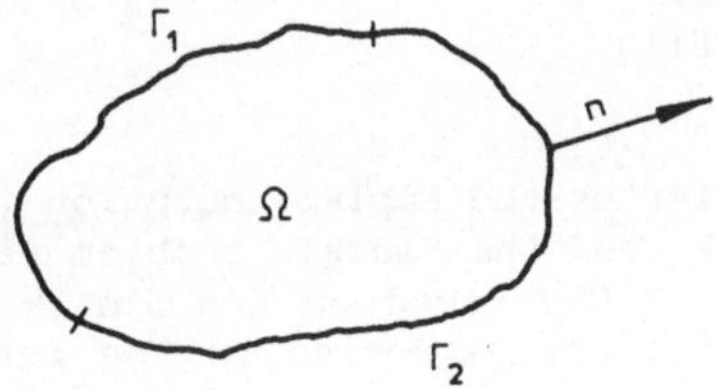

Fig.1

The boundary value problem (1)-(2) seems to be a very "academician" one but we can show two important problems in the fluid mechanics which can be described by similar boundary value problem.

The first one is about an investigation of a potential flow around a profile, according to Fig.2 [PP]. Let S be the given contour of the profile. For a potential flow the continuity equation leads to the existence of a stream function ψ which satisfies the Laplace equation (1). The impenetrability of the contour S means that function ψ on S take a constant value. To satisfy the condition at infinity for $\vec{C}$ one should choose ψ_∞ in a such way that the corresponding vector

$$\vec{C}_\infty = \left(\frac{\partial \psi_\infty}{\partial y}, \frac{\partial \psi_\infty}{\partial x} \right)$$

to be equal to the prescribed constant.

The second example is the problem for an ideal flow in a channel (Fig.3) [PP]. We suppose that at $\pm\infty$ the walls of the channel are parallel with distances D = 1 and **d** respectively. Then the velocities at infinities are known and if we assume that $|C_{-\infty}| = 1$, then $|C_{+\infty}| = \frac{1}{d}$. Then on the walls S_1 and S_2 we get

$\psi|_{S_1} = C$ and $\psi|_{S_2} = C + 1.$

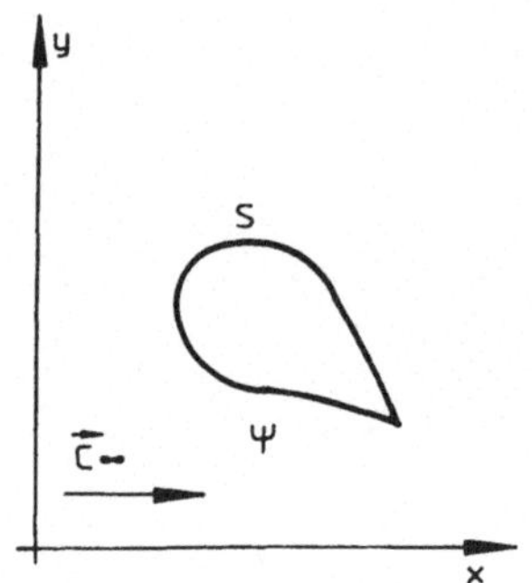

Fig.2 Potential flow around a profile

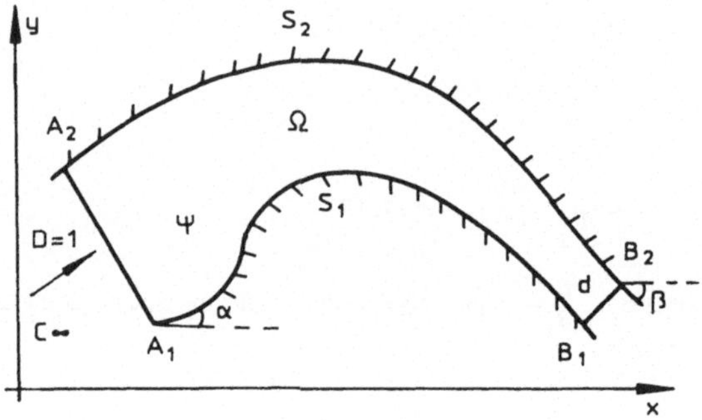

Fig.3 Ideal flow in a channel

The stream function ψ satisfies the Laplace equation (1).

Let's now get back to the model problem (1)-(2). There are two approaches in BEM: "direct" BEM based on Green's formula and "indirect" BEM in which case solutions are expressed in the terms of simple and/or double layer potentials depending on the problem under consideration.

In this paper we will consider "direct" BEM to solve numerically the boundary value problem (1)-(2). It is based on integral equation

$$\alpha\, u(t) = \int_\Gamma q(s)G(s,t)ds - \int_\Gamma u(s)\frac{\partial G(s,t)}{\partial n}\, ds, \quad t \in \bar{\Omega}, \tag{3}$$

where $\alpha = 1$, when $t \in \Omega$ and $\alpha = 0.5$, when $t \in \Gamma$ [BB]. In the equation (3) $G(s,t)$ is the fundamental solution of the Laplacian

$$-\Delta G(s,t) = \delta(s,t), \tag{4}$$

where δ is Dirack delta-function. Suppose a problem in the plane $G(s,t)$ is defined by (5)

$$G(s,t) = \frac{1}{2\pi}\ln\frac{1}{\|s-t\|}, \tag{5}$$

where $\|s-t\| = \left[(s_x - t_x)^2 + (s_y - t_y)^2\right]^{1/2}$.

Let us consider the BEM with constant elements (Fig.4).

The boundary Γ is approximated piecewise linearly and the unknown functions $u(s)$ on Γ_2 and $q(s)$ on Γ_1 are approximated as piecewise constants. Let the boundary Γ_i is divided in N_i boundary elements (i = 1,2) and $N = N_1+N_2$. Let vectors $\underline{u} = \{u_i\}$ and $\underline{q} = \{q_i\}$ consist of values of u(s) and q(s) in the middle points of the boundary elements. We assume an ordering of the nodes such that $\underline{u}^T = \{\underline{u}_1^T, \underline{u}_2^T\}$ and $\underline{q}^T = \{\underline{q}_1^T, \underline{q}_2^T\}$, where the

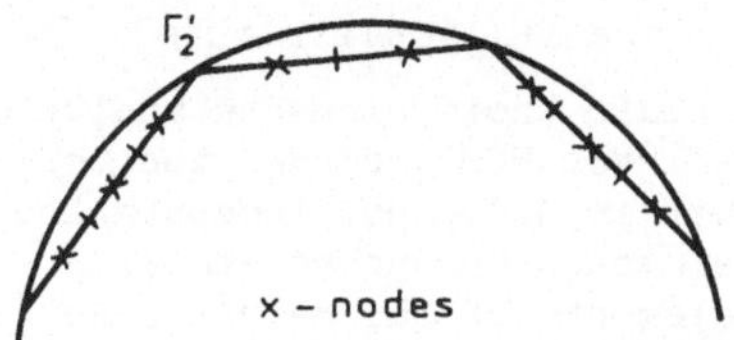

Fig. 4 Constant boundary elements and nodes

unknowns are $\{\underline{u}_2\}$ and $\{\underline{q}_1\}$. The discrete analog of equation (3) is:

$$\begin{bmatrix} G_1 \\ \\ G_2 \end{bmatrix} \begin{Bmatrix} \underline{q}_1 \\ \\ \underline{q}_2 \end{Bmatrix} = \begin{bmatrix} H_1 \\ \\ H_2 \end{bmatrix} \begin{Bmatrix} \underline{u}_1 \\ \\ \underline{u}_2 \end{Bmatrix}, \tag{6}$$

where the elements of the matrices $\mathbf{G}_i$ and $\mathbf{H}_i$ are obtained by formulas:

$$G_i^{kl} = \int_{\Gamma_i^l} G(s_k, t_e) d\Gamma \quad , \quad \hat{H}_i^{kl} = \int_{\Gamma_i^l} \frac{\partial G(s_k, t_e)}{\partial n} d\Gamma. \tag{7}$$

and

$$H_i^{kl} = \begin{cases} \hat{H}_i^{kl} \ , & \text{where } k \neq l \\ \hat{H}_i^{kl} + \frac{1}{2} \ , & \text{where } k = l. \end{cases}$$

To obtain the values of unknowns we have to solve the system of linear algebraic equations

$$A \ \underline{x} = \underline{b}, \tag{8}$$

where

$$A = \begin{bmatrix} G_1 \\ -H_1 \end{bmatrix}, \quad \underline{x} = \{\underline{q}_1^T, \underline{u}_1^T\}, \quad \underline{b} = \begin{bmatrix} H_1 \\ -G_2 \end{bmatrix} \begin{Bmatrix} \underline{u}_1 \\ \underline{q}_2 \end{Bmatrix}. \tag{9}$$

After solving (8) we obtain the nodal unknowns of the function **u** and its first normal derivative **q** on the boundary Γ. Now we are able to calculate value of the function **u** in arbitrary inner point of the domain Ω

and on the boundary Γ. Therefore we use the equation (3) with $\alpha = 1$. A discrete analog of (3) is

$$\underline{u}^* = G^* \underline{q} - H^* \underline{u}, \tag{10}$$

where the vector $\underline{u}^*$ consists of the values of the function **u** in the inner points of Ω and $\mathbf{G}^*$ and $\mathbf{H}^*$ are the corresponding matrices. Their elements are similar to the elements of the matrices $\mathbf{G}_1$ and $\mathbf{H}_1$ from (6). The matrices $\mathbf{G}_1$, $\mathbf{H}_1$, $\mathbf{G}^*$ and $\mathbf{H}^*$ are fully distributed. In the case of constant elements we are able to obtain their elements analytically [BB]. The numerical realization of the direct BEM includes the following steps:

a) initial procedures(input/output, computations of the co-ordinates of the boundary elements and nodes,etc.);
b) computation the elements of the matrix **A** and the right side vector
c) solving the SLAE (8);
d) computation the values of function **u** in the inner points;
e) output operations.

The number of the arithmetic operations in steps b), c) and d) are correspondingly:

$$n_b = 52\, N^2 + O(N);$$

$$n_c = \frac{N}{3}\,(N^2 + 6 \cdot N - 1);$$

$$n_d = 52\, Nm + O(N + m),$$

where m is the number of the inner points and n_c is the number the arithmetic operations for the Gaussian elimination method with partial pivoting (by rows).

A DISTRIBUTED COMPUTER SYSTEM

For the numerical solution of the boundary value problem (1)-(2) by direct BEM we use a distributed computer system. In brief some characteristic of the system are:

- it consist of 4 processors;
- each processor has 1 MB memory;
- there are links between each two processors;
- relatively slow communications between them and between the system and the host (PC XT), one communicative operation equals to about 20 multiplications of real numbers;
- limited Input/Output capabilities (only via one of the processors);
- static configuration of the system (Fig.5);
- high performance processors (T 800 with floating point arithmetic).

This system is a 2D-binary cuboid with communication cost $\mathbf{n \log_2 n - 2}$. In our case n=4. There are the same tasks into the transputers T1, T2 and T3. The task into the root transputer is different. There are all communications with the host computer (I/O operations, operations for the time measure, etc.) in this task. So communications between the transputers are relatively slow we have to find algorithm with as less as possible commu nications. As was mentioned above for solving the SLAE we use Gaussian elimination method with partial pivoting. It is consist of two phases: *forward round* - the matrix is converted into an upper triangular and *backward round* - by successive eliminations the unknowns are obtained. Let the rows with number **i + 4 k,** where k = 0,1,....,[N/4] and the corresponding elements of the right side vector $\underline{b}$ be allocated into the memory of the transputer T_i. Then the computational work for computing of the elements of the matrices G, H, G^* and H^*, the multiplication between the

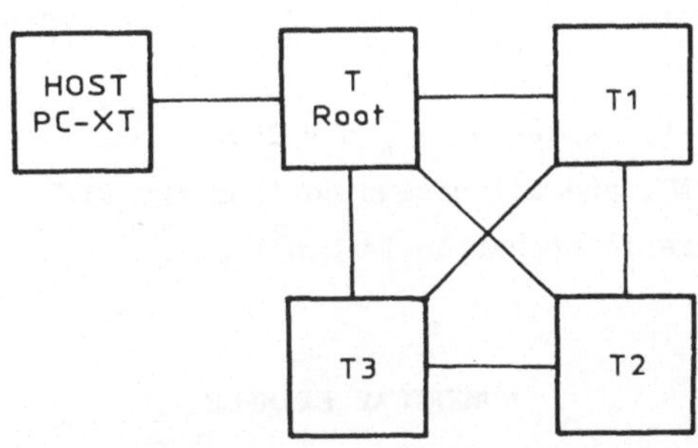

Fig.5 Four transputer system, host and links

matrices G, H, G^* and H^*, the multiplication between the matrices and corresponding vectors and for solving the system of linear equation will be shared between all processors. Each processor performs 1/4 of the operations. To perform this algorithm in parallel the processors have to exchange some intermediate results. In the forward round after **k** steps the processor which contains the row **k** of the matrix A has to send to the others the last **(N - k)** numbers of this row. In the backward round in every step one processor sends to all others **one** number. Thus we need $\mathbf{O(N^2)}$ and $\mathbf{O(N)}$ communicative operations on forward and backward round respectively.

Having in mind this requirement of Gaussian elimination method at the step **b)** the transputer T_i computes only the elements of the rows of the matrix A with number **i + 4 * k** and the same elements of the right side vector **without** any communications.

To obtain the values of the function **u** in some set of interior, for the domain Ω nodes, we have to assemble the matrices $\mathbf{G^*}$ and $\mathbf{H^*}$. After that we have to multiply these matrices with $\underline{q}$ and $\underline{u}$ respectively. Having in mind these operation every transputer computes and put into the memory of the same processor only one quarter of the rows of the matrices $\mathbf{G^*}$ and $\mathbf{H^*}$ without any communications between transputers. As the all elements of the vectors $\underline{q}$ and $\underline{u}$ are allocated into the memory of all processors, this multiplication can be done independently. In this way we obtain a part of the vector $\underline{u}^*$ in every transputer. After that with some communications $\underline{u}^*$ is able to be allocated into the memory of the "root" transputer. Taking into account that there are limited Input/Output capabilities which can be done only via so called "root" transputer we have to read and after that to send as less as possible entry data which every of processors need. For example we read and send only the cartesian co-ordinates of the macro boundary elements and every transputer computes all data connected with them for the boundary elements, nodes, angles, integrals, etc.

Summary, all communications are:

from a step **a)** to a step **b)** - $9N+24 = O(N)$;

from a step **b)** to a step **c)**- **0**;

at a step **c)** - $N*(N-k)+N = O(N^2)+O(N)$;

from a step **c)** to a step **d)** - $6N + \frac{3}{2}m = O(N) + O(m)$;

at a step **d)** - **0**;

from a step **d)** to a step **e)** - $\frac{3}{4}m = O(m)$.

Usually $m = O(N^2)$ and all communications are $O(N^2)$, while the number of arithmetic operation is $O(N^3)$.

NUMERICAL EXAMPLE

The algorithm and codes are tested on the following example.

Let $\Omega = [0.,2.] \times [0.,1.]$ is a rectangle on the plane (Fig.6).Let us consider the boundary value problem (1)-(2) with $u_1(x,y) = x^2-y^2$ and $q_2(x,y) = \frac{\partial u}{\partial n}(x,y)=0$. We take sequencely 100, 200, 300, and 400 boundary elements (half of them on Γ_1 and others on Γ_2).

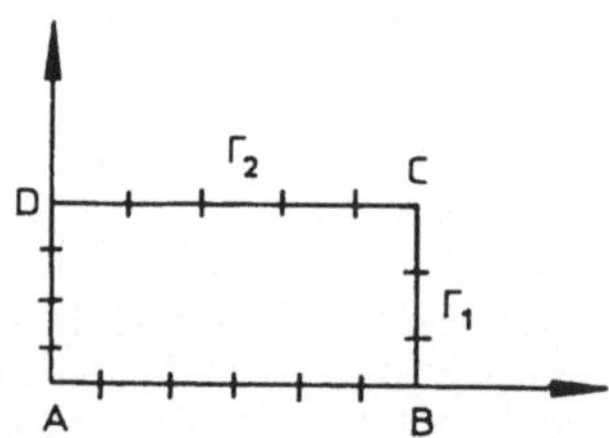

Fig.6 The domain and the boundary of the test example

The results with respect to the execution time and speed-up according to different number of boundary elements and numbers of transputer are shown in the table 1. We have used "3L" operating system and respectively parallel version of Fortran language (near to Fortran 77).

To obtain the speed-up we use the formulae:

$$S_p = \frac{\text{execution time for a single processor}}{\text{execution time using } \mathbf{p} \text{ processors}}.$$

The efficiency of the algorithm $\mathbf{E_p}$ is able to obtain by formula:

$$E_p = \frac{S_p}{p}.$$

In our algorithm we reach an efficiency as following:
- at step **b)** - 0.81;
- at step **c)** - 0.98;
- at step **d)** - 0.81;
- in general - 0.78.

The reached performance is **0.213 Mflops** in case of using only **one** transputer and **0.696 Mflops** when **all four** transputers are used. It is important to note that we have not used more suitable languages for this transputer system as OCAM and C. We have not allocated the data into the register memory as well. One is able to see that in the part of solving the system of linear equations we reach a very good result near to the theoretical one. It is very important because when the number of boundary elements becomes more then 200 the time for solving SLAE's dominates. In another columns the volumes include the time for the communications between transputers at the previous step. The execution of almost the same codes on PC XT takes about 15 minutes when the number of boundary elements is 100.

Table 1. Execution time and speed-up according to different number of processors and boundary elements

N of processors	N of boundary elem.	Time for initial precedures	Time for computation of **A** and **b**	Time for solution of **A** $\underline{x} = \underline{b}$	Time for solution in inner points	Total time in sec.
1	100	11.59	3.86	4.76	8.07	45.58
4	100	10.06	1.19	1.46	2.49	35.03
Speed-up			**3.24**	**3.25**	**3.24**	**1.42**
1	200	20.34	15.51	37.83	16.17	106.82
4	200	18.32	4.78	10.29	4.98	58.07
Speed-up			**3.24**	**3.68**	**3.24**	**2.17**
1	300	30.97	34.96	129.32	36.39	261.10
4	300	26.98	10.77	33.56	11.22	109.54
Speed-up			**3.25**	**3.85**	**3.24**	**2.78**
1	400	39.91	62.18	305.38	64.72	508.35
4	400	35.42	19.16	77.77	19.96	186.32
Speed-up			**3.25**	**3.93**	**3.24**	**3.11**

REFERENCES:

[BB] Banerjee P., Butterfield R., *Boundary Element Method in Engineering Science*, McGraw-Hill, 1981

[BK] Boyanov K., Kanchev T.,et al., *Computing System with Parallel Processing*, Technica, 1986 (in Bulgarian)

[PP] Panov L., Pasheva V, Lazarov R., Lubenov V., *Flows with free lines*, Proc. of Summer School for Young Scientists, 1976, Varna, Bulgaria

Direct Evaluation of Hypersingular Integrals in 2D BEM

M. Guiggiani
Dipartimento di Costruzioni Meccaniche e Nucleari
Università degli Studi di Pisa
via Diotisalvi 2, 56126 Pisa, Italy

Summary

A general direct method for the evaluation of hypersingular integrals in the BEM is presented with reference to two-dimensional problems. It is first shown that there are no special problems in the derivation of hypersingular boundary integral equations (HBIE's), that is, in taking the singular point on the boundary. No unbounded quantities ultimately arise if the limiting process is properly performed. In the second part, the limit is expressed in terms of intrinsic coordinates. This is maybe the most critical step since the effects of the mapping must be accounted for. Then, the use of suitable expansions allows all singular integration to be performed analytically, even if curved higher-order boundary elements are employed. The original hypersingular integral is thus shown to be equivalent to some simple computable terms.

Introduction

Hypersingular boundary integral equations (HBIE's) are attracting increasing attention as also shown by the growing number of papers devoted to this issue (e.g., [1, 2, 3, 7, 8, 10, 11, 12, 13, 14, 15], to mention but a few).

In most cases, integrals with hypersingular kernels are regularized *before* any numerical treatment. Integration by parts [1, 2, 12, 13, 15], Stokes' theorem [10], or simple solutions [14] are employed to obtain integral equations free of hypersingular kernels. Other authors proposed more direct analytical methods ([3]). However, pure analytical techniques are limited to *flat* surfaces.

A different general approach to the numerical treatment of hypersingular boundary integral equations was proposed by Guiggiani et al. in [7] and [8] for three-dimensional scalar and vector problems, respectively.

First it was shown that no unbounded terms actually arise when dealing with HBIE's, provided the limit is properly taken. The limiting process was then "translated" in terms of intrinsic coordinates. The use of suitable expansions allowed all singular integrals to be evaluated analytically, and the limit to be carried out exactly. The remaining regular integrals were evaluated numerically. The simplicity of the final formula allowed hypersingular integrals to be evaluated even on *curved* boundary elements. Standard Gaussian quadrature formulae were used in numerical tests.

In this paper, the same approach is presented for two-dimensional problems. The basic steps are the same as for 3D applications. However, some differences make the 2D case worthy of its own presentation.

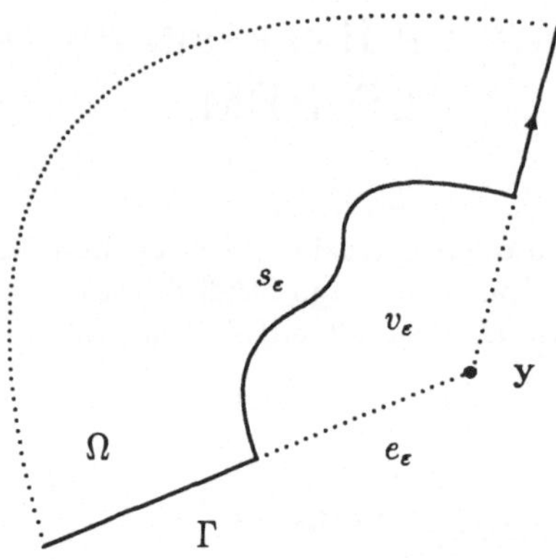

Figure 1: Exclusion of the singular point $\mathbf{y}$ by a vanishing neighbourhood v_ε.

Although this paper deals mainly with potential problems, the proposed approach is by no means limited to this field. It is applicable in any other field and with any kind of boundary elements.

Hypersingular Boundary Integral Equations

In this section the limiting process for the derivation of hypersingular boundary integral equations will be outlined. It will be shown that no *a priori* "interpretations", either in the Cauchy Principal Value (CPV) or in the Finite Part (FP) sense, are really necessary. If the singular point is taken to the boundary through a proper limiting process, no unbounded terms ultimately arise. Moreover, a more general formulation is obtained.

For brevity, potential problems are considered (Laplace operator). However, the method is entirely applicable to any other scalar or vector problem (cf. [7] and [8]).

Let us first consider the standard boundary integral equation for the harmonic function $u(\mathbf{x})$ on a two-dimensional domain Ω bounded by the regular (in the sense of Kellog [9]) curve Γ with unit outward normal $\vec{n}(\mathbf{x}) = \{n_i\}$ (Figure 1)

$$\lim_{\varepsilon \to 0^+} \left\{ \int_{(\Gamma - e_\varepsilon) + s_\varepsilon} [T(\mathbf{y}, \mathbf{x}) u(\mathbf{x}) - U(\mathbf{y}, \mathbf{x}) q(\mathbf{x})] ds_x \right\} = 0, \tag{1}$$

where $U = -(1/2\pi) \ln r$ is the fundamental solution of Laplace equation, $q = \partial u/\partial n$, and $T = \partial U/\partial n$. If r denotes $|\mathbf{x} - \mathbf{y}|$, that is the distance between the generic (integration) point $\mathbf{x} = \{x_i\}$ and the singular point $\mathbf{y} = \{y_i\}$, we have $U = O(\ln r)$ (weak singularity) and $T = O(r^{-1})$ (strong singularity), when $r \to 0$. On the other hand, U is of class C^∞ for each pair $\mathbf{x}$, $\mathbf{y}$ with $\mathbf{x} \neq \mathbf{y}$. These properties are a particular case of the general properties of the fundamental solutions of elliptic operators with constant coefficients (e.g., [16], p. 154).

Due to the singularities in U and T, a small region v_ε around $\mathbf{y} \in \Gamma$ was removed from Ω. Notice that it is *not* necessary to restrict the shape of v_ε. It may have any shape, provided $\mathbf{y}$ is an exterior point for $\Omega_\varepsilon = \Omega - v_\varepsilon$, and the boundary $\Gamma_\varepsilon = (\Gamma - e_\varepsilon) + s_\varepsilon$

of Ω_ε is a regular curve [9] (Figure 1). Equation (1) is the Green's second identity for the two harmonic functions u and U on Ω_ε.

The identity holds even in the limit, that is when v_ε shrinks down on $\mathbf{y}$ (we assume that the maximum chord of v_ε approaches 0 when $\varepsilon \to 0$). The overall limit is not affected by the shape of v_ε. Notice that, only in the very special case of v_ε being part of a circle centered at $\mathbf{y}$, the line integral in (1) will match the CPV definition which requires e_ε to be *symmetric* around $\mathbf{y}$.

Since both functions $U(\mathbf{y},\mathbf{x})$ and $T(\mathbf{y},\mathbf{x})$ are of class C^∞ in Ω_ε, we can differentiate under the integral sign with respect to any of the coordinates y_k of the singular point, thus obtaining

$$\lim_{\varepsilon\to 0^+}\left\{\int_{(\Gamma-e_\varepsilon)+s_\varepsilon}[V_i(\mathbf{y},\mathbf{x})u(\mathbf{x})-W_i(\mathbf{y},\mathbf{x})q(\mathbf{x})]ds_x\right\}=0, \tag{2}$$

where $W_i = \partial U/\partial y_i$ and $V_i = \partial T/\partial y_i$.

For 2D potential problems, we have

$$\begin{aligned} W_i(\mathbf{y},\mathbf{x}) &= \frac{1}{2\pi r}r_{,i}\ , \\ V_i(\mathbf{y},\mathbf{x}) &= -\frac{1}{2\pi r^2}\left[2r_{,i}\frac{\partial r}{\partial n}-n_i(\mathbf{x})\right], \end{aligned} \tag{3}$$

where $r_{,i} = \partial r/\partial x_i = -\partial r/\partial y_i$. The kernel W_i shows a *strong* singularity of order r^{-1} while the kernel V_i is *hypersingular* of order r^{-2}, as $r \to 0$.

Now, we assume that the potential $u \in C^{1,\alpha}$ at $\mathbf{y}$, that is, u is continuous with its first partial derivatives at $\mathbf{y} \in \Gamma$, and all first derivatives satisfy a Hölder condition. Consequently, the following expansions hold (repeated indices imply summation)

$$\begin{aligned} u(\mathbf{x}) &= u(\mathbf{y}) + u_{,k}(\mathbf{y})(x_k - y_k) + O(r^{1+\alpha}), \\ q(\mathbf{x}) = u_{,k}(\mathbf{x})n_k(\mathbf{x}) &= u_{,k}(\mathbf{y})n_k(\mathbf{x}) + O(r^\alpha), \end{aligned} \tag{4}$$

where α is a positive constant (usually, $\alpha = 1$). It is worth noting that the discretization scheme has to satisfy the same regularity requirements at each collocation point.

Expansions (4) allow the identity (2) to become

$$\begin{aligned} &\lim_{\varepsilon\to 0^+}\left\{\int_{(\Gamma-e_\varepsilon)}[V_i(\mathbf{y},\mathbf{x})u(\mathbf{x})-W_i(\mathbf{y},\mathbf{x})q(\mathbf{x})]ds_x\right. \\ &+\int_{s_\varepsilon}\Big(V_i[u(\mathbf{x})-u(\mathbf{y})-u_{,k}(\mathbf{y})(x_k-y_k)]-W_i[q(\mathbf{x})-u_{,k}(\mathbf{y})n_k(\mathbf{x})]\Big)ds_x \\ &\left.+u(\mathbf{y})\int_{s_\varepsilon}V_i\,ds_x+u_{,k}(\mathbf{y})\int_{s_\varepsilon}[V_i(x_k-y_k)-W_i\,n_k(\mathbf{x})]ds_x\right\}=0. \end{aligned} \tag{5}$$

At this stage it is convenient to take s_ε in the shape of a *circular* arc. However, this choice does not affect the overall final result.

Owing to the simple shape of s_ε, all integrals on s_ε in (5) can be evaluated explicitly, thus obtaining

$$\lim_{\varepsilon\to 0^+}\int_{s_\varepsilon}\Big(V_i[u(\mathbf{x})-u(\mathbf{y})-u_{,k}(\mathbf{y})(x_k-y_k)]-W_i[q(\mathbf{x})-u_{,k}(\mathbf{y})n_k(\mathbf{x})]\Big)ds_x=0, \tag{6}$$

so that we only have to consider the limit of the other integrals on s_ε. They are given by (see Appendix for a detailed derivation when $\mathbf{y}$ is at a corner point)

$$\lim_{\varepsilon\to 0^+}\int_{s_\varepsilon}[V_i(x_k-y_k)-W_i n_k(\mathbf{x})]ds_x = c_{ik}(\mathbf{y}), \tag{7}$$

and

$$\lim_{\varepsilon\to 0^+}\int_{s_\varepsilon} V_i ds_x = \lim_{\varepsilon\to 0^+}\frac{b_i(\mathbf{y})}{\varepsilon}, \tag{8}$$

where c_{ik} and b_i are (bounded) coefficients. They depend upon the local geometry of Γ at $\mathbf{y}$. (besides the selected shape of s_ε)

At smooth boundary points the free-term coefficients of the hypersingular boundary integral equation simply become $c_{ik}(\mathbf{y}) = 0.5\,\delta_{ik}$.

By collecting the previous results, the *hypersingular* boundary integral equation can be written in the following form

$$c_{ik}(\mathbf{y})u_{,k}(\mathbf{y}) + \lim_{\varepsilon\to 0^+}\left\{\int_{(\Gamma-e_\varepsilon)}[V_i(\mathbf{y},\mathbf{x})u(\mathbf{x}) - W_i(\mathbf{y},\mathbf{x})q(\mathbf{x})]\,ds_x + u(\mathbf{y})\frac{b_i(\mathbf{y})}{\varepsilon}\right\} = 0. \tag{9}$$

A similar identity holds for vector problems [8].

In equation (9) the limiting process is still indicated explicitly. We already computed c_{ik} and b_i as limits of integrals on s_ε, with s_ε being a circular arc. Consistently, in (9) e_ε must be a *symmetric* neighbourhood of $\mathbf{y}$.

Notice that the finite-part integral concept plays no role in the derivation of (9). We needed no special "interpretations" at all, since all terms are meaningful even in the classical framework. Moreover, it was shown that restrictions in the shape of v_ε are not necessary. They are merely convenient to make analytical manipulations easier.

Instead of regularizing the singular kernels in (9) by either integration by parts ([15, 13, 1, 12, 2]) or Stokes' Theorem ([10]), we proceed directly to the numerical treatment. Of course, special care has to be exercised to properly "translate" the limiting process in terms of intrinsic coordinates when the discretization for the geometry is introduced.

Direct Evaluation of Hypersingular Integrals: 2D Case

Let Γ_s be a part of the contour Γ containing the singular point $\mathbf{y}$. In general, in two-dimensional BEM analyses the collocation point $\mathbf{y}$ may be either shared by two boundary elements (Figure 2-a), or it may lie (strictly) within one boundary element (Figure 2-b). The natural choice in the first case is $\Gamma_s = \Gamma_s^1 \cup \Gamma_s^2$, whereas in the second case Γ_s is just the element containing $\mathbf{y}$.

The main numerical issue is the evaluation of the *bounded* quantity

$$\lim_{\varepsilon\to 0^+}\left\{\int_{(\Gamma_s-e_\varepsilon)} V_i(\mathbf{y},\mathbf{x})u(\mathbf{x})ds_x + u(\mathbf{y})\frac{b_i(\mathbf{y})}{\varepsilon}\right\}. \tag{10}$$

with hypersingular kernel $V_i = O(r^{-2})$. Consistently with the already evaluated coefficients c_{ik} and b_i, the neighbourhood e_ε must be symmetric, that is $e_\varepsilon = \{\mathbf{x}\in\Gamma \mid |\mathbf{x}-\mathbf{y}| \le \varepsilon\}$.

On each boundary element, the potential u is represented by shape functions $N^c(\xi)$ of local intrinsic coordinate $\xi \in [-1,\ 1]$. Usually, $u = u[\mathbf{x}(\xi)] = \sum_c N^c(\xi)\,a^c$ results in a polynomial representation (a^c may or may not be nodal values). Notice that it is $u(\mathbf{x})$ that must be $C^{1,\alpha}$ at $\mathbf{y}$, not $u[\mathbf{x}(\xi)]$.

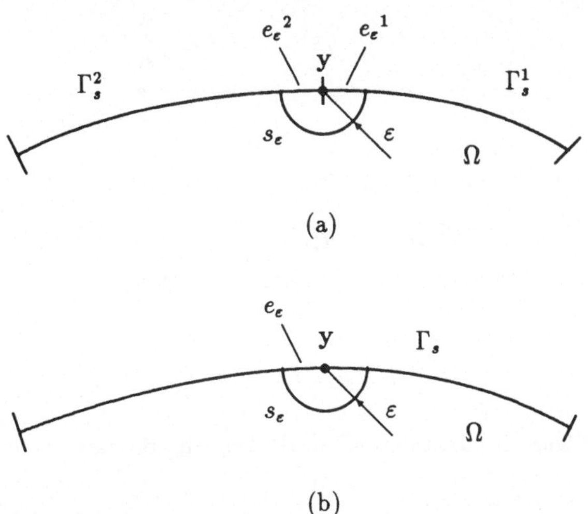

Figure 2: Limiting process on the contour Γ: (a) pole $\mathbf{y}$ between two boundary elements; (b) pole $\mathbf{y}$ within one boundary element.

Limiting process and discretization of the geometry

Let us consider in detail the case of $\mathbf{y}$ belonging to two boundary elements, say Γ_s^1, Γ_s^2 (Figure 2-a). By means of the usual representation for the geometry in terms of shape functions and nodal coordinates

$$x_k(\xi) = \sum_c M^c(\xi)\, x_k^c, \qquad k = 1, 2, \tag{11}$$

each boundary element is mapped onto the interval $\xi \in [-1,\ 1]$. We denote by R_s^m the interval on the ξ-axis associated to the boundary element Γ_s^m. The elementary arc length is given by $ds_x = J^m(\xi)\, d\xi$.

Accordingly, the two parts $e_\varepsilon{}^1$ and $e_\varepsilon{}^2$ of the symmetric neighbourhood e_ε are mapped, respectively, onto two portions $\sigma_\varepsilon{}^1$ and $\sigma_\varepsilon{}^2$ of the corresponding intervals R_s^m. It is very important to note that, in general, $\sigma_\varepsilon{}^1$ and $\sigma_\varepsilon{}^2$ have different length. We denote the length of each $\sigma_\varepsilon{}^m$ by $\alpha_m(\varepsilon)$, $m = 1, 2$ (Figure 3).

We are now ready to state the problem of the evaluation of integrals with hypersingular kernel in terms of intrinsic coordinate, which is what we actually need in a BEM framework. We can provide the following equivalent expressions

$$\begin{aligned} I &= \lim_{\varepsilon\to 0^+} \left\{ \sum_{m=1}^{2} \int_{\Gamma_s^m - e_\varepsilon{}^m} V_i(\mathbf{y},\mathbf{x}) N^c(\xi(\mathbf{x}))\, ds_x + N^c(\eta) \frac{b_i(\mathbf{y})}{\varepsilon} \right\} \\ &= \lim_{\varepsilon\to 0^+} \left\{ \sum_{m=1}^{2} \int_{R_s^m - \sigma_\varepsilon{}^m} V_i(\mathbf{y},\mathbf{x}(\xi)) N^c(\xi) J^m(\xi)\, d\xi + N^c(\eta) \frac{b_i(\mathbf{y})}{\varepsilon} \right\} \\ &= \lim_{\varepsilon\to 0^+} \left\{ \int_{-1}^{1-\alpha_1(\varepsilon)} F^1(1,\xi)\, d\xi + \int_{-1+\alpha_2(\varepsilon)}^{1} F^2(-1,\xi)\, d\xi + N^c(\eta) \frac{b_i(\mathbf{y})}{\varepsilon} \right\}, \end{aligned} \tag{12}$$

where $F^m(\eta,\xi) = V_i\, N^c\, J^m$, and η is the image of $\mathbf{y}$ on each interval R_s^m ($\eta = 1$ for $m = 1$, and $\eta = -1$ for $m = 2$) (Figure 3).

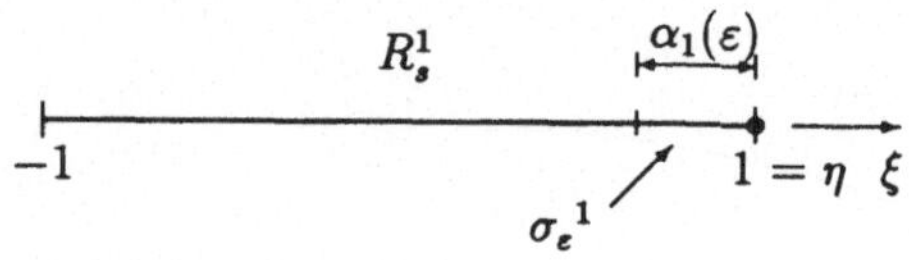

Figure 3: "Translation" of the limiting process.

Series expansions

In this subsection we want to derive the following expansions

$$F^m(\eta,\xi) = \frac{F_{-2}^m}{(\xi-\eta)^2} + \frac{F_{-1}^m}{\xi-\eta} + O(1), \tag{13}$$

and

$$\rho = \alpha_m(\varepsilon) = \varepsilon\,\beta_m + \varepsilon^2\,\gamma_m\ \mathrm{sgn}(\delta) + O(\varepsilon^3), \tag{14}$$

where $\delta = \xi - \eta$, $\rho = |\delta|$, and $\mathrm{sgn}(\delta) = |\delta|/\delta$. The quantities β_m and γ_m are *constants* typical of the mapping between the element Γ_s^m and the interval R_s^m. Similarly, F_{-2}^m and F_{-1}^m are *constants* that depend on the local characteristic of $F^m(\eta,\xi)$ at $\xi = \eta$.

Before providing a systematic way to obtain F_{-2}^m, F_{-1}^m, β_m and γ_m, some more basic expansions are derived.

Everything is actually based on the Taylor expansion of $(x_i - y_i)$

$$\begin{aligned} x_i - y_i &= \left.\frac{dx_i}{d\xi}\right|_{\xi=\eta}(\xi-\eta) + \left.\frac{d^2x_i}{d\xi^2}\right|_{\xi=\eta}\frac{(\xi-\eta)^2}{2} + \cdots \\ &= A_i\,(\xi-\eta) + B_i\,(\xi-\eta)^2 + \cdots \\ &= A_i\,\delta + B_i\,\delta^2 + O(\delta^3), \end{aligned} \tag{15}$$

where, as indicated, all derivatives are evaluated at η. Therefore, A_i and B_i are just constants. They can be easily obtained from (11) for *any kind* of boundary element.

It is now convenient to define

$$A = \left(\sum_{k=1}^{2} {A_k}^2\right)^{1/2} = J(\eta) > 0, \tag{16}$$

$$B = \left(\sum_{k=1}^{2} {B_k}^2\right)^{1/2} \geq 0, \tag{17}$$

$$C = \sum_{k=1}^{2} A_k B_k. \tag{18}$$

Notice that $s_i = A_i/A$ are the components of the unit tangent vector to Γ at $\mathbf{y}$.

From the definition of r and accordingly to the previous results, the powers of r are given by

$$r^n = A^n|\delta|^n\left(1+n\frac{C}{A^2}\delta\right)+O(\delta^{(n+2)}). \tag{19}$$

Also useful are the expasions for the derivatives $r_{,i} = \partial r/\partial x_i$

$$\begin{aligned} r_{,i} &= \frac{x_i-y_i}{r} = \frac{A_i\,\delta + B_i\,\delta^2 + O(\delta^3)}{A|\delta|\left(1+\frac{C}{A^2}\delta\right)+O(\delta^3)} \\ &= \operatorname{sgn}(\delta)\frac{A_i}{A} + \left(\frac{B_i}{A}-\frac{A_iC}{A^3}\right)|\delta| + O(\delta^2), \end{aligned} \tag{20}$$

and for the inverse powers of r

$$\frac{1}{r^2} = \frac{1}{A^2\delta^2} - \frac{2C}{A^4\delta} + O(1). \tag{21}$$

In general, the expansions for all inverse powers of r can be obtained directly from (19) with negative n.

The Jacobian is given by $J(\xi) = \sqrt{J_1^2(\xi)+J_2^2(\xi)} = \sqrt{(dx_1/d\xi)^2+(dx_2/d\xi)^2}$. It is useful to have the expansions of $J_i = n_i\,J = J_{i0} + J_{i1}\,\delta + O(\delta^2)$ (the suffix m is dropped)

$$J_1 = A_2 + 2B_2\delta + O(\delta^2), \tag{22}$$

$$J_2 = -A_1 - 2B_1\delta + O(\delta^2). \tag{23}$$

For completeness, the expansion for $J(\xi)$ is also provided

$$J(\xi) = J(\eta) + \left.\frac{dJ}{d\xi}\right|_{\xi=\eta}(\xi-\eta) + \cdots = A + \frac{2C}{A}\delta + O(\delta^2). \tag{24}$$

The last basic expansion we need is for the generic shape function

$$N^c(\xi) = N^c(\eta) + \left.\frac{dN^c}{d\xi}\right|_{\xi=\eta}(\xi-\eta) + \cdots = N_0^c + N_1^c\delta + O(\delta^2). \tag{25}$$

It is now easy to obtain the constants β_m and γ_m in the expansion (14) for $\alpha_m(\varepsilon)$. First observe that on each part $e_\varepsilon{}^m$ of the vanishing neighbourhood on Γ_s the distance between $\mathbf{x}$ and $\mathbf{y}$ is equal to ε (Figure 2)

$$r = \varepsilon. \tag{26}$$

Therefore, by taking the series expansion for r (cf. (19))

$$r = A|\delta| + \frac{C}{A}|\delta|\delta + O(\delta^3) = \varepsilon, \tag{27}$$

and using the reversion of the above series (Figure 3)

$$\begin{aligned} \rho = \alpha_m(\varepsilon) &= \frac{\varepsilon}{A} - \varepsilon^2\frac{C}{A^4}\operatorname{sgn}(\delta) + O(\varepsilon^3) \\ &= \varepsilon\,\beta_m + \varepsilon^2\gamma_m\operatorname{sgn}(\delta) + O(\varepsilon^3), \end{aligned} \tag{28}$$

where $\rho = |\delta|$. Expression (28) shows that $\beta_m = 1/A$, and $\gamma_m = -C/A^4$. It is worth noting that they have the same formal expressions as for three-dimensional problems (see Appendix C in [7]).

The explicit expressions of F^m_{-2} and F^m_{-1} can be obtained by using the same systematic approach as described in [7] for three-dimensional problems. It is just a matter of taking the expression of $F^m(\eta,\xi) = V_i N^c J^m$, and replacing each term by its expansion (remember that $n_i J = J_i$). For two-dimensional potential problems, whose hypersingular kernels are given in (3), the required expressions are as follows (some drastic simplifications occur)

$$F^m_{-2} = \frac{1}{2\pi}\frac{N^c_0 J_{i0}}{A^2} = \frac{1}{2\pi}\frac{n_i(\mathbf{y})N^c(\eta)}{J^m(\eta)}, \tag{29}$$

$$F^m_{-1} = \frac{1}{2\pi}\frac{N^c_1 J_{i0}}{A^2} = \frac{1}{2\pi}n_i(\mathbf{y})\left.\frac{dN^c}{ds}\right|_{x=y} = \frac{1}{2\pi}\frac{n_i(\mathbf{y})}{J^m(\eta)}\left.\frac{dN^c}{d\xi}\right|_{\xi=\eta}. \tag{30}$$

Expressions (29) and (30) define explicitly the first two terms in the series expansion (13) of $F^m(\eta,\xi)$.

We see from (30) that F^m_{-1} is the same on both elements ($F^1_{-1} = F^2_{-1}$) when $\mathbf{y}$ is at a *smooth* boundary point. Indeed, $n_i(\mathbf{y})$ is the same on both boundary elements. Moreover, we supposed $u \in C^{1,\alpha}$ at $\mathbf{y}$. Therefore, we had to employ shape functions such that $dN^c/ds \in C^{0,\alpha}$ at $\mathbf{y}$. Notice that the Jacobian may well be different between the two elements (therefore, in general, $dN/d\xi$ is *not* continuous).

Owing to the possible discontinuity of the Jacobian, F^m_{-2} is not the same (in general) on both elements.

So far we considered the pole $\mathbf{y}$ to be shared by two elements. However, the whole analysis in this subsection is entirely applicable to the case of $\mathbf{y}$ strictly within one element (Figure 2-b). Obviously, in this case the values of F_{-2} and F_{-1} are the same on both sides of $\mathbf{y}$.

Analytical treatment of hypersingular integrals

We are now ready for the actual evaluation of I as given by the last limit in (12). First, let us add and subtract in each integral the first two terms of the series expansion (13)

$$\begin{aligned} I &= \lim_{\varepsilon\to 0^+}\left\{\sum_{m=1}^{2}\int_{R^m_s-\sigma_\varepsilon{}^m}\left(F^m(\eta,\xi) - \left[\frac{F^m_{-2}}{(\xi-\eta)^2} + \frac{F^m_{-1}}{\xi-\eta}\right]\right)d\xi \right.\\ &\left. \quad + \int_{R^m_s-\sigma_\varepsilon{}^m}\frac{F^m_{-1}}{\xi-\eta}\,d\xi + \left(\int_{R^m_s-\sigma_\varepsilon{}^m}\frac{F^m_{-2}}{(\xi-\eta)^2}\,d\xi + N^c(\eta)\frac{b_i(\mathbf{y})}{\varepsilon}\right)\right\} \\ &= I_0 + I_{-1} + I_{-2}\,. \end{aligned} \tag{31}$$

Let us consider each term separately. From (13) we see that in I_0 the integrand is regular. Therefore the limit is straightforward

$$I_0 = \int_{-1}^{1}\left\{\left(F^1(1,\xi) - \left[\frac{F^1_{-2}}{(\xi-1)^2} + \frac{F^1_{-1}}{\xi-1}\right)\right] + \left(F^2(-1,\xi) - \left[\frac{F^2_{-2}}{(\xi+1)^2} + \frac{F^2_{-1}}{\xi+1}\right)\right]\right\}d\xi\,. \tag{32}$$

This is regular integral that can be numerically evaluated by standard formulae.

I_{-1} and I_{-2} have very simple integrands. Therefore, they are integrable in *closed form*. Moreover, thanks to the expansions (14) for $\alpha_m(\varepsilon)$, the limiting process can be performed *exactly*.

For I_{-1} we have

$$I_{-1} = \lim_{\varepsilon\to 0^+}\left\{\int_{-1}^{1-\alpha_1(\varepsilon)}\frac{F^1_{-1}}{\xi-1}\,d\xi + \int_{-1+\alpha_2(\varepsilon)}^{1}\frac{F^2_{-1}}{\xi+1}\,d\xi\right\}$$

$$
\begin{aligned}
&= \lim_{\varepsilon\to 0^+}\left\{F^1_{-1}\ln\left|\frac{-\alpha_1(\varepsilon)}{-2}\right| + F^2_{-1}\ln\left|\frac{2}{\alpha_2(\varepsilon)}\right|\right\} \\
&= F_{-1}\lim_{\varepsilon\to 0^+}\left\{\ln\left|\frac{\varepsilon\beta_1}{2}\right| + \ln\left|\frac{2}{\varepsilon\beta_2}\right|\right\} \\
&= F_{-1}\lim_{\varepsilon\to 0^+}\{\ln|\varepsilon| + \ln|\beta_1| - \ln|2| + \ln|2| - \ln|\varepsilon| - \ln|\beta_2|\} \\
&= F_{-1}\ln\left|\frac{\beta_1}{\beta_2}\right|. \qquad (33)
\end{aligned}
$$

where $F_{-1} = F^1_{-1} = F^2_{-1}$. It is worth remembering that $\beta_m = 1/J^m(\eta)$.
An equivalent expression for I_{-1} is

$$
I_{-1} = F_{-1}\sum_{m=1}^{2}\left(\operatorname{sgn}(\delta)\ln\left|\frac{2}{\beta_m}\right|\right), \qquad (34)
$$

which is formally similar to the corresponding term in the 3D case (cf. (28) in [7]).

The treatment for I_{-2} is similar. We first integrate analytically, and then perform the limiting process using the expansions for $\alpha_m(\varepsilon)$. Notice that the unbounded terms arising from the integration are cancelled by the term $N^c(\eta)b_i(\mathbf{y})/\varepsilon$

$$
\begin{aligned}
I_{-2} &= \lim_{\varepsilon\to 0^+}\left\{\int_{-1}^{1-\alpha_1(\varepsilon)}\frac{F^1_{-2}}{(\xi-1)^2}\,d\xi + \int_{-1+\alpha_2(\varepsilon)}^{1}\frac{F^2_{-2}}{(\xi+1)^2}\,d\xi + N^c(\eta)\frac{b_i(\mathbf{y})}{\varepsilon}\right\} \\
&= \lim_{\varepsilon\to 0^+}\left\{F^1_{-2}\left(\frac{1}{\alpha_1(\varepsilon)} - \frac{1}{2}\right) + F^2_{-2}\left(\frac{1}{\alpha_2(\varepsilon)} - \frac{1}{2}\right) + N^c(\eta)\frac{b_i(\mathbf{y})}{\varepsilon}\right\} \\
&= \lim_{\varepsilon\to 0^+}\left\{\sum_{m=1}^{2}\left[\frac{F^m_{-2}}{\varepsilon\beta_m}\left(1 - \frac{\varepsilon\,\gamma_m\operatorname{sgn}(\delta)}{\beta_m}\right)\right] + N^c(\eta)\frac{b_i(\mathbf{y})}{\varepsilon}\right\} - \frac{1}{2}\sum_{m=1}^{2}F^m_{-2} \\
&= \lim_{\varepsilon\to 0^+}\left(\frac{1}{\varepsilon}\right)\left\{\sum_{m=1}^{2}\frac{F^m_{-2}}{\beta_m} + N^c(\eta)b_i(\mathbf{y})\right\} - \sum_{m=1}^{2}\left[F^m_{-2}\left(\frac{\gamma_m\operatorname{sgn}(\delta)}{(\beta_m)^2} + \frac{1}{2}\right)\right] \\
&= -\sum_{m=1}^{2}\left[F^m_{-2}\left(\frac{\gamma_m\operatorname{sgn}(\delta)}{(\beta_m)^2} + \frac{1}{2}\right)\right]. \qquad (35)
\end{aligned}
$$

This formula is equivalent to expression (26) in [7] (3D case). In particular $\operatorname{sgn}(\delta)\gamma_m$ corresponds to $\gamma(\theta)$, since $\gamma(\theta) = -\gamma(\theta+\pi)$.

In the above expressions, the "local terms" β_m and γ_m account for the *distorsion* of the originally symmetric neighbourhood as introduced by the geometric discretization of the boundary (Figures 2 and 3).

Final formula for hypersingular integrals

It is worth noting that I_0 has been reduced to a regular integral, whereas I_{-1} and I_{-2} are given by bounded constants. Therefore, the quantity I—initially given by integrals of hypersingular functions plus some unbounded terms (cf. (31))—is now easily computable.

Indeed, it is just a matter of collecting the previous results. By adding expressions (33), (34), and (35), the following formula for hypersingular integrals in 2D BEM analyses is obtained

$$
I = \sum_{m=1}^{2}\left\{\int_{-1}^{1}\left[F^m(\eta,\xi) - \left(\frac{F^m_{-2}}{(\xi-\eta)^2} + \frac{F^m_{-1}}{\xi-\eta}\right)\right]d\xi\right.
$$

$$+ F_{-1}^m \ln\left|\frac{2}{\beta_m}\right| \mathrm{sgn}(\xi-\eta) - F_{-2}^m\left[\mathrm{sgn}(\xi-\eta)\frac{\gamma_m}{(\beta_m)^2}+\frac{1}{2}\right]\Bigg\}, \tag{36}$$

where $\eta = 1$ for $m = 1$, and $\eta = -1$ for $m = 2$. The numbers 2 are the lengths of the ξ-intervals $(-1 \leq \xi \leq 1)$.

Formula (36) holds for the case of $\mathbf{y}$ being shared by two boundary elements (Figure 2-a). If the collocation point lies within only one element (Figure 2-b) the corresponding formula is as follows

$$\begin{aligned} I &= \int_{-1}^{1}\left[F(\eta,\xi)-\left(\frac{F_{-2}}{(\xi-\eta)^2}+\frac{F_{-1}}{\xi-\eta}\right)\right]d\xi \\ &\quad + F_{-1}\ln\left|\frac{1-\eta}{\eta+1}\right| - F_{-2}\left(\frac{1}{1-\eta}+\frac{1}{\eta+1}\right). \end{aligned} \tag{37}$$

In this case the lengths of the two subintervals are equal to $\eta+1$ and $1-\eta$, respectively. Since all quantities A_k, B_k are the same on both sides of η, there are no local terms.

Notice that both formulae (36) and (37) are *exact*. No approximations have been introduced.

Formulae (36) and (37) have full generality. They holds for boundary elements of any kind and order. Following the same steps, similar expressions can be derived for any other elliptic operator (e.g., elasticity, acoustics, plate bending, elastodynamics, etc.). The only difference is in the expressions of F_{-2} and F_{-1}. However, for any field they can be obtained by means of the basic expansions (15)–(25). Incidentally, it is worth noting that expressions (29) and (30) for F_{-2} and F_{-1} also hold for steady-state acoustics, since the corresponding kernels have the same asymptotic behaviour. Similarly, elastostatic and elastodynamic kernels have the same asymptotic behaviour [1, 4].

Concluding Remarks

In some recent papers ([7, 8]) a new direct method for the evaluation of hypersingular integrals in the BEM was proposed with reference to three-dimensional problems. In this paper, the same approach has been described for two-dimensional BEM analyses.

Although emphasis here has been on potential problems, the proposed approach is completely general. Not only it can be applied in any field, but there are no restrictions on the selection of boundary elements. Therefore, higher-order curved boundary elements can be easily employed.

It is also worth noting that the proposed direct approach provides a unified method for the treatment of singular integrals in the BEM. Indeed, the formula for strongly singular integrals (i.e., Cauchy principal values) as given in [5] is just a special case of formula (36) with $F_{-2} = 0$. The same happens in 3D problems (cf. [6] and [7, 8]). Therefore, we may even conjecture that a hierarchy of formualae like (36) could be obtained for the evaluation of integrals with arbitrarily high order of singularity.

Acknowledgements

Work performed with the support of MURST and CNR.

References

[1] Bonnet M., 1989, "Regular boundary integral equations for three-dimensional finite or infinite bodies with or without curved cracks in elastodynamics", *Boundary Element Techniques: Applications in Engineering*, Brebbia C.A. and Zamani N., eds., Computational Mechanics Publications, Southampton, pp. 171–188..

[2] Cruse T. A., and Novati G., 1990, "Traction BIE formulations and applications to non-planar and multiple cracks", forthcoming.

[3] Gray L. J., Martha L. F., and Ingraffea A. R., 1990, "Hypersingular integrals in boundary element fracture analysis", *Int. J. Num. Methods Eng.*, Vol. 29, pp. 1135–1158.

[4] Guiggiani M., 1991, "Computing principal value integrals in 3D BEM for time-harmonic elastodynamics—A direct approach", *Comm. in Appl. Numerical Methods*, forthcoming.

[5] Guiggiani M., and Casalini P., 1987, "Direct computation of Cauchy principal value integrals in advanced boundary elements", *Int. J. Num. Methods Eng.*, Vol. 24, pp. 1711–1720.

[6] Guiggiani M., and Gigante A., 1990, "A general algorithm for multidimensional Cauchy principal value integrals in the boundary element method", ASME *Journal of Applied Mechanics*, Vol. 57, pp. 906–915.

[7] Guiggiani M., Krishnasamy G., Rudolphi T. J., and Rizzo F. J., 1991, "A general algorithm for the numerical solution of hypersingular boundary integral equations", ASME *Journal of Applied Mechanics*, (to appear).

[8] Guiggiani M, Krishnasamy G., Rizzo F. J., and Rudolphi T. J., 1991, "Hypersingular boundary integral equations: A new approach to their numerical treatment", *Proc. IABEM-90 Conference*, October 15–19, 1990, Rome, Italy, Springer-Verlag (in press).

[9] Kellog O. D., 1929, *Foundations of Potential Theory*, Frederick Ungar Publishing Comp., New York.

[10] Krishnasamy G., Schmerr L. W., Rudolphi T. J., and Rizzo F. J., 1990, "Hypersingular boundary integral equations: Some applications in acoustic and elastic wave scattering", ASME *Journal of Applied Mechanics*, Vol. 57, pp. 404–414.

[11] Martin P. A., and Rizzo F. J., 1989, "On boundary integral equations for crack problems", *Proc. Royal Soc. London*, Vol. A 421, pp. 341–355.

[12] Nishimura N., and Kobayashi S., 1989, "A regularized boundary integral equation method for elastodynamic crack problems", *Computational Mechanics*, Vol. 4, pp. 319–328.

[13] Polch E. Z., Cruse T. A., and Huang C.-J., 1987, "Traction BIE solutions for flat cracks", *Computational Mechanics*, Vol. 2, pp. 253–267.

[14] Rudolphi T. J., 1991, "The use of simple solutions in the regularization of hypersingular boundary integral equations", *Mathematical and Computer Modelling*, Special Issue on BIEM/BEM, in press.

[15] Sládek V., and Sládek J., 1984, "Transient elastodynamic three-dimensional problems in cracked bodies", *Applied Mathematical Modelling*, Vol. 8, pp. 2–10.

[16] Villaggio P., 1977, *Qualitative Methods in Elasticity*, Noordhoff International Publishing, Leyden.

Appendix: Coefficients c_{ik} and b_i for Hypersingular Boundary Integral Equations

In this Appendix, the free-term coefficients c_{ik} and b_i associated to hypersingular boundary integral equations for two-dimensional potential problems are derived. They are defined by expressions (7) and (8), respectively.

s_ε was assumed to be a circle centered at $\mathbf{y}$ and of radius ε. The angle with respect to a horizontal axis is denoted by φ, so that, on s_ε, $\varphi_1 \leq \varphi \leq \varphi_2$. Owing to the selected shape of s_ε, the following simple relations hold when $\mathbf{x} \in s_\varepsilon$

$$\begin{aligned} &r = \varepsilon, && \partial r/\partial n = -1, \\ &r_{,1} = \cos\varphi, && r_{,2} = \sin\varphi, \\ &n_1 = -\cos\varphi, && n_2 = -\sin\varphi, \\ &x_1 - y_1 = \varepsilon \cos\varphi, && x_2 - y_2 = \varepsilon \sin\varphi. \end{aligned} \tag{A1}$$

and $ds_x = \varepsilon\, d\varphi$. Moreover, the kernel functions become

$$\begin{aligned} W_1 &= \frac{\cos\varphi}{2\pi\varepsilon}, & W_2 &= \frac{\sin\varphi}{2\pi\varepsilon}, \\ V_1 &= \frac{\cos\varphi}{2\pi\varepsilon^2} = \frac{W_1}{\varepsilon}, & V_2 &= \frac{\sin\varphi}{2\pi\varepsilon^2} = \frac{W_2}{\varepsilon}. \end{aligned} \tag{A2}$$

The integrations in (7) and (8) are now a straightforward matter. For instance

$$\frac{b_1(\mathbf{y})}{\varepsilon} = \int_{s_\varepsilon} V_1\, ds_x = \frac{1}{2\pi}\int_{\varphi_1}^{\varphi_2} \frac{\cos\varphi}{\varepsilon^2}\, \varepsilon\, d\varphi = \frac{\sin\varphi_2 - \sin\varphi_1}{2\pi}\,\frac{1}{\varepsilon}. \tag{A3}$$

Note that $b_1 = 0$ if $\varphi_2 = \varphi_1 + 2\pi$, that is if $\mathbf{y}$ is an internal point. Similar observations hold for b_2 which is given by

$$b_2 = \frac{\cos\varphi_1 - \cos\varphi_2}{2\pi}. \tag{A4}$$

However, these quantities do not need to be evaluated in actual computations.

An analogous treatment applies for coefficients c_{ik}. For instance

$$\begin{aligned} c_{11}(\mathbf{y}) &= \int_{s_\varepsilon} [V_1(x_1 - y_1) - W_1 n_1]\, ds_x = \frac{1}{2\pi}\int_{\varphi_1}^{\varphi_2} 2\cos\varphi\, d\varphi \\ &= \frac{1}{2\pi}\left[(\varphi_2 - \varphi_1) + \frac{\sin(2\varphi_2) - \sin(2\varphi_1)}{2}\right]. \end{aligned} \tag{A5}$$

Note that $c_{11} = 0$ if $\mathbf{y}$ is an internal point, while $c_{11} = 0.5$ if $\mathbf{y}$ is at a smooth boundary point, i.e., if $\varphi_2 = \varphi_1 + \pi$. The other coefficients are given by

$$c_{22} = \frac{1}{2\pi}\left[(\varphi_2 - \varphi_1) - \frac{\sin(2\varphi_2) - \sin(2\varphi_1)}{2}\right], \tag{A6}$$

and

$$c_{12} = c_{21} = \frac{\sin^2\varphi_2 - \sin^2\varphi_1}{2\pi}. \tag{A7}$$

At smooth boundary points, $c_{22} = 0.5$, and $c_{12} = c_{21} = 0$, as expected.

EFFICIENT ALGORITHMS FOR VECTOR OR PARALLEL-COMPUTING TO ANALYZE NONLINEAR GRAVITY WAVES WITH THE BOUNDARY ELEMENT METHOD

C. Haack, V. Schlegel, O. Mahrenholtz
Meerestechnik II, TU Hamburg-Harburg
Eißendorfer Str. 42, D-2100 Hamburg 90
FR Germany

SUMMARY

The numerical simulation of gravity waves with nonlinear free surfaces needs — even if the boundary element method is used — a large number of elements. In particular this is valid for the analysis of three-dimensional problems. To perform extensive numerical experiments efficient algorithms are needed, taking advantage of the facilities of modern vector- or parallel-computers. Different solution schemes for the boundary integral equations as the point-collocation, the Bubnov-Galerkin method and a least squares method as well as different equation solvers are tested for their possibilities of vectorization. In order to quantify this potential the influence of the vectorization on the computing time of the numerical simulation is shown. The accuracy of the different vectorized solution schemes is demonstrated by comparing their numerical results for a regular periodic deep-water wave with the analytical solution for this problem.

INTRODUCTION

For many ocean engineering applications an efficient and accurate calculation scheme is necessary to model and analyze the time-dependent extremely nonlinear behavior of ocean waves. The boundary element method is well suited for this type of problems because for such large domains with nonlinear boundaries full advantage of the efficient discretization procedure of this method can be taken. The approach we use — a combined boundary element time stepping technique — was introduced by Longuet-Higgins and Cokelet [4] (1976). Although this well known and reliable numerical method for the simulation of nonlinear gravity waves was optimized and extended by several other authors, even nowadays computation-time is the most restrictive factor for practical purposes.

Consequently, modern computer architecture concepts like vector or parallel-processing have to be exploited in order to reduce run time. Therefore, the main CPU-time dependencies of the combined boundary element time stepping method are pointed out and the concept of vectorization for different solution schemes is explained.

PROBLEM FORMULATION

COMBINED BOUNDARY ELEMENT TIME STEPPING TECHNIQUE

In order to describe the free surface potential flow, the boundaries Γ_1 to Γ_4 of the domain Ω are discretized by finite boundary elements as shown in Fig. 1.

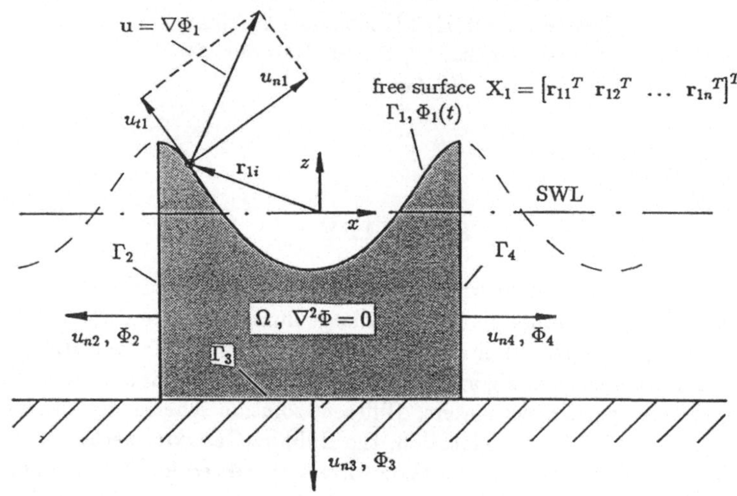

Figure 1: Free surface potential flow

If an incompressible fluid and an irrotational flow is assumed, the Laplace equation is valid in the fluid domain,

$$\nabla^2\Phi = 0 \ . \tag{1}$$

By means of a direct boundary element method equation (1) is transformed into a Cauchy integral equation of the form

$$C(\boldsymbol{\xi})\,\Phi(\boldsymbol{\xi}) = \int\!\!\!\!-_{\Gamma} \left(\frac{\partial\Phi(\mathbf{x})}{\partial \mathbf{n}(\mathbf{x})} G(\mathbf{x},\boldsymbol{\xi}) - \Phi(\mathbf{x}) \frac{\partial G(\mathbf{x},\boldsymbol{\xi})}{\partial \mathbf{n}(\mathbf{x})} \Phi \right) d\Gamma(\mathbf{x}) \ , \tag{2}$$

with $\Gamma = \Gamma_1 + \Gamma_2 + \Gamma_3 + \Gamma_4$. The position vectors for field and source points are denoted by $\mathbf{x}$ and $\boldsymbol{\xi}$ and the unit outward normal vector by $\mathbf{n}$. $C(\boldsymbol{\xi})$ is a geometric constant and $G(\mathbf{x},\boldsymbol{\xi})$ the fundamental solution of the plane problem . On the free surface Γ_1 the potential Φ_1 is given as an initial value problem

$$\Phi_1 = \Phi_1(t) \ , \tag{3}$$

and on the sea bottom Γ_3 the normal velocity vanishes, see Fig. 1:

$$u_{n3} = 0 \ . \tag{4}$$

In the nonlinear theory, along the vertical boundaries Γ_2 and Γ_4 no other condition than the periodicity is known:

$$u_{n2} = -u_{n4} \,, \tag{5}$$

$$\Phi_2 = \Phi_4 \quad . \tag{6}$$

With the given boundary conditions for the fluid the Cauchy integral equation (2) results in a system of linear equations of the form:

$$\mathbf{A}(\mathbf{X}_1)\ \mathbf{a} = \mathbf{b}(\mathbf{X}_1) \ , \tag{7}$$

where **A**, the matrix of coefficients and **b**, the vector of known boundary conditions, depend on the position vector $\mathbf{X}_1$ of the fluid particles on the free surface (see Fig. 1). The vector of unknown boundary conditions is denoted by **a** with

$$\mathbf{a}^T = \left[u_{n1}{}^T\ \Phi_2{}^T\ \Phi_3{}^T\ u_{n2}{}^T\right]^T \quad . \tag{8}$$

For a given potential $\Phi_1(t)$ on Γ_1 we can solve the Laplace equation (1) and derive the unknown velocity field on the free surface:

$$\mathbf{u} \ = \ [u_n \ u_t]^T \ = \ \nabla\Phi \ . \tag{9}$$

The tangential component u_t of the velocity vector $\mathbf{u}$ can be calculated, e.g. by spline interpolation of the given potential Φ_1.

In order to describe the motion of the fluid particles on the free surface, in a next step we have to solve an initial value problem by integrating the time-dependent linear kinematic and nonlinear dynamic boundary conditions

$$\frac{d\mathbf{X}_1}{dt} = \nabla\Phi_1 \,, \tag{10}$$

$$\frac{d\Phi_1}{dt} = -\frac{1}{\rho}p - g\,(\mathbf{z}_1 - \overline{\mathbf{z}}_1) + \frac{1}{2}(\nabla\Phi_1)^2 \,. \tag{11}$$

Here t denotes the time, p the pressure, ρ the density of the fluid, g the acceleration due to gravity, $\mathbf{z}_1$ the vertical reference coordinate of the fluid particles on the free surface and $\overline{\mathbf{z}}_1$ the undisturbed surface — the still water line (SWL). These coupled ordinary differential equations (10) and (11) are solved by using an Adams-Bashforth-Moulton time integration scheme. Thus, we obtain the nonlinear boundary conditions on the free surface for the next time step $i + 1$. Because the matrix of coefficients **A** depends on the position vector $\mathbf{X}_1$, the Cauchy integral equation (2) has to be solved for each time step. Figure 2 shows the simplified structure of the algorithm for the proposed combined boundary element time stepping technique.

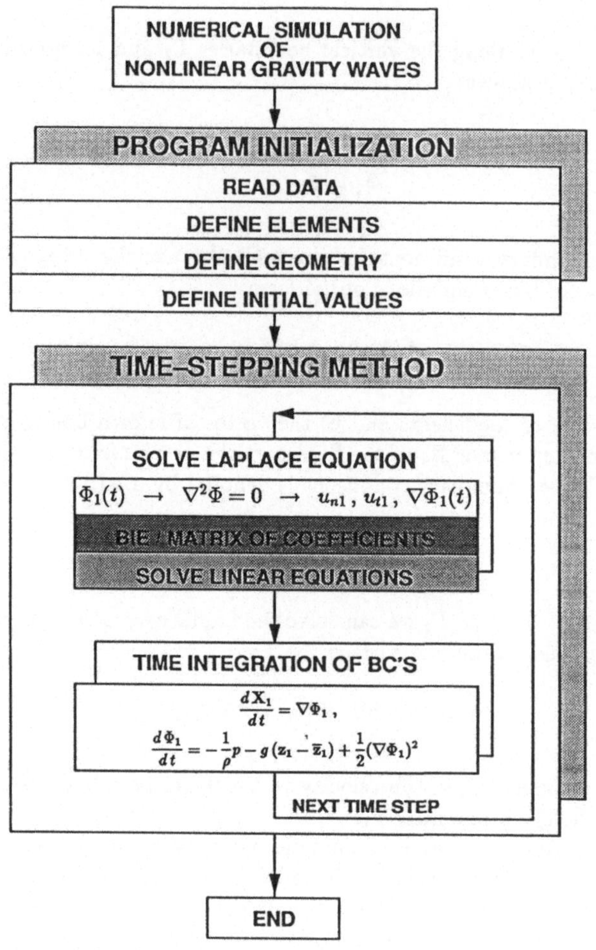

Figure 2: Flow-chart of the algorithm for the combined boundary element time stepping technique

ASSEMBLY OF THE MATRIX OF COEFFICIENTS

The different solution schemes for the boundary integral equation (2) can be classified by the way they minimize the defect

$$\mathbf{D}(\boldsymbol{\xi}) = \fint_{\Gamma} \left(\frac{\partial \mathrm{w}(\mathbf{x},\boldsymbol{\xi})}{\partial \mathbf{n}(\mathbf{x})} \, \Phi(\mathbf{x}) - \mathrm{w}(\mathbf{x},\boldsymbol{\xi}) \, \frac{\partial \Phi(\mathbf{x})}{\partial \mathbf{n}(\mathbf{x})} \right) \, d\Gamma(\mathbf{x}) - \mathrm{C}\,\Phi(\mathbf{x}) \; . \qquad (12)$$

Weighted residual schemes are working with different weight functions $\mathrm{w}(\boldsymbol{\xi})$, with

$$\int_{\Gamma} \mathbf{D}(\boldsymbol{\xi})\, \mathrm{w}(\boldsymbol{\xi})\, d\Gamma(\boldsymbol{\xi}) = 0 \;, \tag{13}$$

whereas least squares methods try to minimize the quadratic form

$$\int_{\Gamma} \mathbf{D}^2(\boldsymbol{\xi})\, d\Gamma(\boldsymbol{\xi}) \Rightarrow \text{Min.} \;. \tag{14}$$

Method A – point-collocation: using the Dirac-function δ as test function in equation (13)

$$\mathrm{w}(\boldsymbol{\xi}) = \delta(\boldsymbol{\xi} - \boldsymbol{\xi}_j) \;, \tag{15}$$

this point-collocation results directly in a simple system of linear equations due to the special properties of the Dirac-function

$$\mathbf{A}\,\mathbf{a} = \mathbf{b} \;. \tag{16}$$

In this case the matrix of coefficients $\mathbf{A}$ becomes a dense and unsymmetric matrix.

Method B – Galerkin: An additional numerical integration is required to solve the integral in equation (13) if polynomial Ansatz-functions φ_k^T are chosen as test functions $\mathrm{w}(\boldsymbol{\xi})$:

$$\sum_{j=1}^{n_{GP}} \mathbf{D}(\boldsymbol{\xi}_j)\, \varphi(\mathbf{x}_{kj})\, \mathbf{Q} = \mathbf{Q}\mathbf{D} = 0 \;, \tag{17}$$

with n_{GP}, the number of grid points $\boldsymbol{\xi}$. This classical Bubnov-Galerkin approach yields linear equations of the form

$$\mathbf{Q}\,\mathbf{A}\,\mathbf{a} = \mathbf{Q}\,\mathbf{b} \;, \tag{18}$$

where $\mathbf{Q}$ is a diagonal matrix of the integration weights multiplied by the determinant of the Jacobian matrix. Here, in the case of mixed boundary conditions, i.e. Dirichlet- and Neumann-conditions, the matrix of coefficients $\mathbf{Q}\,\mathbf{A}$ is a dense and unsymmetric matrix.

Method C – least squares: minimizing the quadratic form of equation (14) requires an additional numerical integration as well

$$\sum_{j=1}^{n_{GP}} \mathbf{D}^2(\boldsymbol{\xi}_j)\, \mathbf{Q} = \mathbf{D}^T\, \mathbf{Q}\, \mathbf{D} \Rightarrow \text{Min.} \;. \tag{19}$$

Thus, the weighted Gaussian normal equations are obtained:

$$\underbrace{\mathbf{A}^T\,\mathbf{Q}\,\mathbf{A}}_{\text{symmetric}}\; \mathbf{a} = \mathbf{A}^T\,\mathbf{Q}\,\mathbf{b} \,, \tag{20}$$

with the now symmetric matrix $\mathbf{A}^T \mathbf{Q} \mathbf{A}$. For the Dirichlet-condition on the free surface this procedure achieves the highest order of convergence, see Table 1. The order of convergence and the condition for polynomial Ansatz-functions of order $k-1$ given by Wendland [7] are listed in Table 1 for the different solution schemes for boundary integral equations (BIE's). For Neumann-conditions the factor 2α has to be zero and for Dirichlet-conditions this factor has to be set to -1.

Table 1: Order of convergence and condition of different solution schemes for BIE's

method	point-collocation	Galerkin	least squares
denotation	**A**	**B**	**C**
convergence	$\mathcal{O}(h^{k-2\alpha})$	$\mathcal{O}(h^{2k-2\alpha})$	$\mathcal{O}(h^{2k-4\alpha})$
condition	$\mathcal{O}(h^{-\|2\alpha\|})$	$\mathcal{O}(h^{-\|2\alpha\|})$	$\mathcal{O}(h^{-\|4\alpha\|})$

SOLUTION SCHEMES FOR LINEAR EQUATIONS

For each of the introduced methods A, B and C two different procedures for the solution of linear equations were investigated. To show the effect of already optimized programs on CPU-time reduction some of them were taken from a vectorized program-library (VECLIB). Table 2 shows the different solution schemes which were chosen for methods A, B and C.

Table 2: Solution schemes for linear equations

denotation	used for method	solution scheme
S1	**A, B**	Schmidt-orthonormalization
S2	**A, B**	QR-Householder (using VECLIB)
S3	**C**	Cholesky decomposition (using symmetric storage mode)
S4	**C**	Cholesky decomposition (full stored matrix, VECLIB)

CPU-TIME ANALYSIS

The computation time for the combined boundary element time stepping technique (see Fig. 2) depends mainly on the number of time steps, the number of boundary elements, the chosen solution methods and on the computer hardware. All measured CPU-times we give are valid for the vector-machine CONVEX C220.

A strictly linear CPU-time dependency of the number of time steps was observed, whereas the number of elements has an exponential influence on the computation time. Therefore, an upper limit of about 100 elements was given for practical calculations. With the help of modern programming tools a CPU-time analysis was accomplished to detect the most time consuming parts of the program. The main result is shown in Fig. 3. For the methods A,B and C we observed a CPU-time consumption of 90% for generating the matrix of coefficients (MATGEN) and the solution of the linear equations

(ESOLVE). The same result is given by Romate [5] for the point-collocation, method A. For a small number of elements the MATGEN-part has a great influence on the CPU-time, but with increasing number of elements the influence of the ESOLVE-part becomes more important, see Fig. 3. Compared to the other methods the least squares method (C) requires significantly more computation time due to the extra matrix operations. On the other hand the symmetric properties of the matrix of coefficients can be exploited. This results in less computation time for the Cholesky decomposition scheme, see Fig. 3.

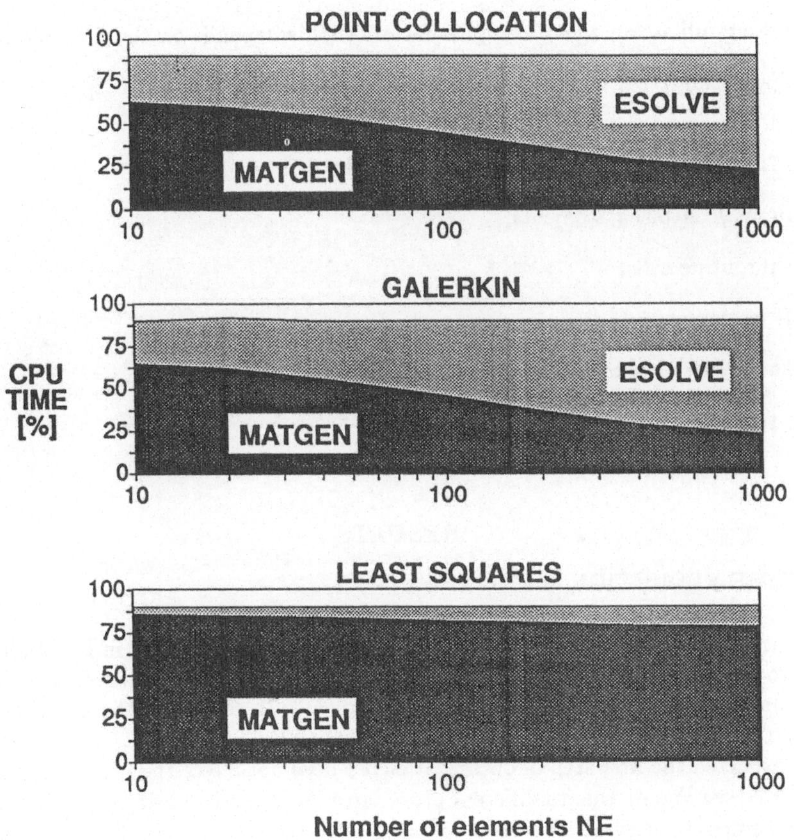

Figure 3: CPU-time analysis

VECTORIZATION

The degree of optimization and consequently the CPU-time reduction depends on the vector length or on the matrix dimension and the internal machine organization. However, the vectorization is a first step towards parallel-processing and is already realized in a practical and efficient way. The principles of different computer architecture concepts are pointed out in detail e.g. by Schendel [6]. Normally it is not possible to affect the internal machine organization, but with the CONVEX-System we have the ability to optimize the program-code (here FORTRAN) with respect to the vectorization of basic vector, vector-matrix and matrix-matrix operations. We did this in two steps:

Step V1:

- using already optimized programs for basic linear algebraic operations (BLAS-routines), see Coleman and Van Loan [1],
- taking full advantage of compiler facilities (this depends on the machine type),
- optimizing DO-LOOPS in the program-code, see e.g. Gentzsch [3].

Step V2: optimizing the program structure by avoiding

- input / output statements,
- subroutine calls,
- IF-statements,
- and recursive calculations

within DO-LOOPS.

RESULTS

CPU-TIME REDUCTION

The results of the proposed two degrees of vectorization are presented in Fig. 4 and Fig. 5, where the CPU-time consumption of the different program segments MATGEN, ESOLVE and of the complete program (MAIN) is given with respect to the non-optimized program version (V0) and the number of elements (NE).

As expected the first step of optimization V1 does not affect the matrix generation MATGEN, see Fig. 4. Due to different procedures for numerical integration of singular and non-singular integrals the vector-structure within the program is destroyed and the vector facilities can not be exploited for the assembly of the matrix of coefficients. This is independent of the chosen method A,B or C. The use of already optimized library-routines (VECLIB) for the QR-Householder (ESOLVE, S2) and the Cholesky decomposition scheme (ESOLVE, S4), results even for a full stored matrix in a maximal CPU-time reduction of approximately 90%, see Fig. 4. As pointed out before this behavior extremely dependends on the vector length, i.e. the number of elements NE. Using these schemes for the solution of linear equations the CPU-time requirement of the complete program is reduced to less than 50% of the primary version, if point-collocation (MAIN, AS2) or the Galerkin-method (MAIN, BS2) is used.

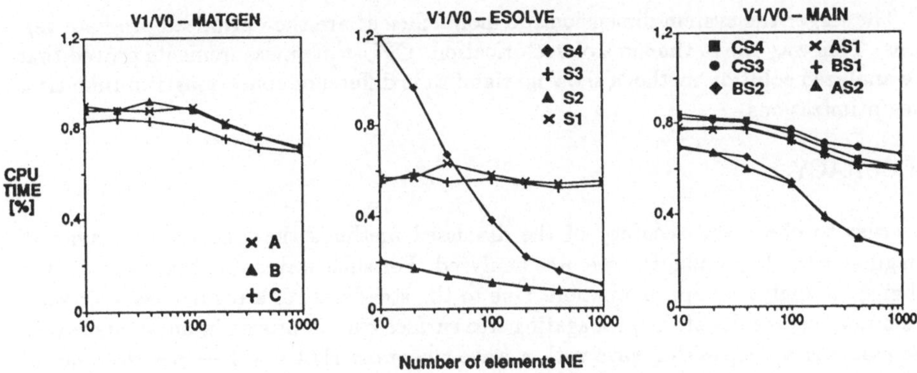

Figure 4: CPU-time reduction due to the first step of vectorization V1

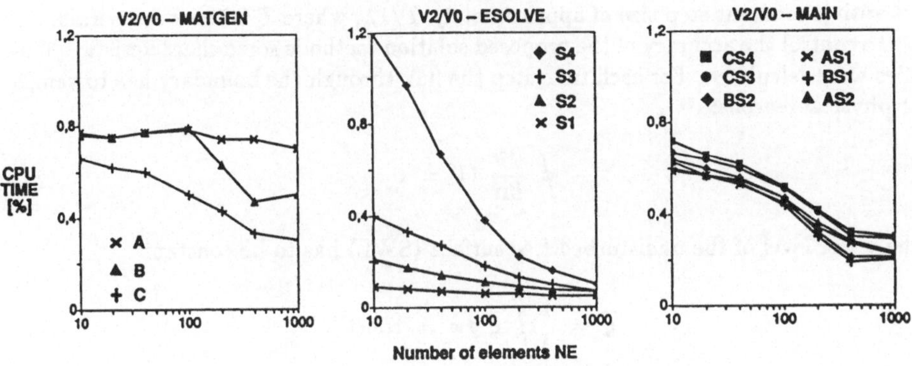

Figure 5: CPU-time reduction due to the second step of vectorization V2

The second degree of vectorization V2 can minimize the CPU-time consumption of the least squares method (C) to 50–30% of the non-optimized program V0, see MATGEN Fig. 5 . Compared to methods A and B, now the additional matrix operations in equation (20) were exploited. Furthermore, this second step results in an extreme CPU-time reduction for the solution of linear equations, ESOLVE. By using the symmetric storage mode the time requirement of the Cholesky decomposition scheme was reduced to less than 7% of the primary version V0. This agrees to a speed up factor of 15. Independent of the chosen solution methods the second step of optimization achieves a CPU-time reduction of 50–80% for the combined boundary element time stepping method under consideration, see MAIN Fig. 5.

In the existing version the handling of numerical integration procedures for the solution of boundary integral equations leads to restrictions due to vectorization. Here parallel-processing is the most efficient approach to solve this generell problem, see e.g. Zucchini and Mukherjee [8].

The larger the system dimension the more efficient are the optimized program versions. This was one of the aims of vectorization. CPU-time measurements proved that the analyzed solution methods show no significant difference concerning run time after the optimizations.

ACCURACY

In order to check the accuracy of the discussed methods the long time behavior of a regular periodic nonlinear wave was analyzed. For such waves Rienecker and Fenton [2] give an analytical approximation. Due to the steadiness of a regular periodic wave the numerical results of the propagating free surface can be proved in an accurate way. We assumed a deep water wave with a wave steepness $H/\lambda = 0.1$ — the wave height is denoted by H, the wave length by λ — and a fraction $d/\lambda = 1$, where d is the water depth. One wave length λ of the free surface is discretized into 50 cubic spline elements. Two polygonial elements of fourth order are used for each of the vertical boundaries and three elements for the sea bottom. The time integration was carried out with a constant step size of approximately $T/12$, where T is the wave period.

To control the accuracy of the proposed solution methods some characteristic quantities were calculated. For each time step the flux through the boundary has to vanish for physical reasons:

$$\oint \frac{\partial \Phi}{\partial n} \, d\Gamma \overset{!}{=} 0 . \tag{21}$$

The water level of the undisturbed free surface (SWL) has to be constant

$$\overline{\zeta} = \frac{1}{\lambda} \int_0^\lambda \zeta \, dx \overset{!}{=} \text{const.}, \tag{22}$$

where ζ is the elevation of the fluid particles with respect to the undisturbed surface and x is the horizontal coordinate. In addition, the total energy of the considered system has to be constant in time

$$\mathbf{E} = \mathbf{E}_{\text{kin}} + \mathbf{E}_{\text{pot}} = \frac{\rho}{2} \int_\Gamma \Phi \frac{\partial \Phi}{\partial n} \, d\Gamma + \frac{\rho}{2} \int_0^\lambda \zeta^2 \, dx \overset{!}{=} \text{const.}, \tag{23}$$

where $\mathbf{E}_{\text{kin}}$ is the kinetic and $\mathbf{E}_{\text{pot}}$ the potential energy.

For this application the results of the numerical simulations were independent of the chosen solution scheme for solving linear equations. Even with more than 10 periods T the long time behavior shows acceptable results and no numerical instabilities occur. Furthermore, for all analyzed methods the relative error concerning the still water line comes out to be $10^{-3}\%$, independent of the simulation time. Figure 6 depicts the error of the flux through the boundaries with regard to time t for the analyzed methods A, B and C. Even in this case no difference is observed between the Galerkin (B) and the least squares method (C). Compared to point-collocation (A) these methods result in a more stable behavior. In Fig. 7 the error of the total energy is given. At the beginning of the numerical simulation the error grows significantly but for longer times an asymptotical behavior is observed.

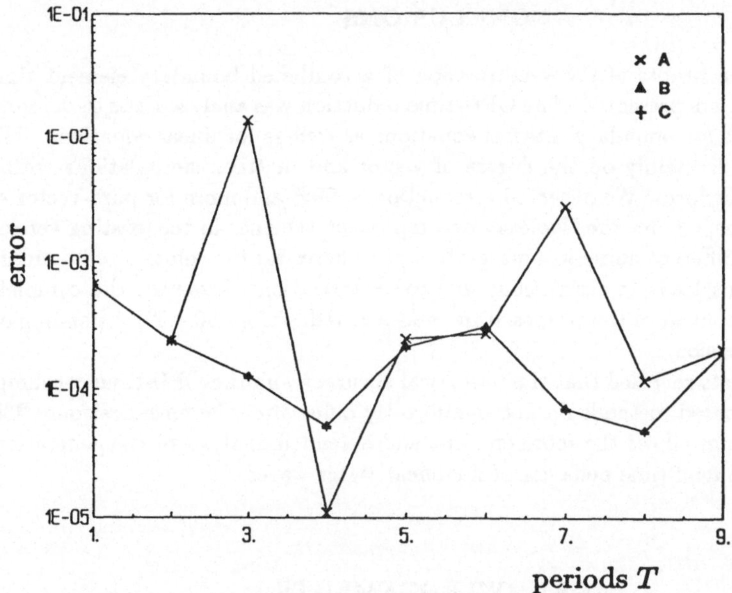

Figure 6: Error concerning the flux through the boundaries

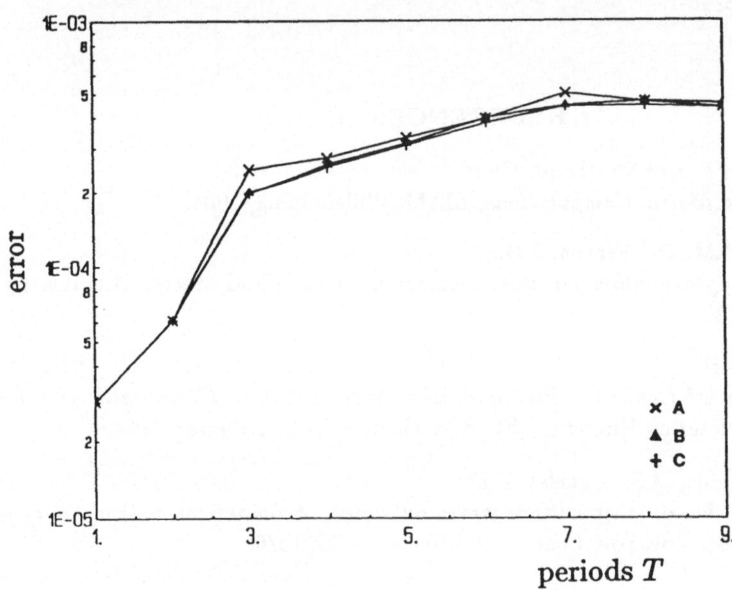

Figure 7: Error concerning the total energy

CONCLUSIONS

In this paper the results of the vectorization of a combined boundary element time stepping method are presented. The CPU-time reduction was analyzed due to different solution methods for boundary integral equations as well as for linear equations. The reduction depends mainly on the degree of vector and matrix manipulations within the solution procedures. We observed a reduction of 90% and more for pure vector or matrix operations, e.g. for the Cholesky decomposition scheme. In the existing version the handling of different numerical integration procedures for the solution of boundary integral equations leads to restrictions due to vectorization. However, the complete CPU-time requirement of the proposed method was reduced by 50–80% depending on the system dimension.

Numerical tests revealed that the numerical accuracy and the CPU-time consumption of the compared methods do not significantly differ after the optimizations. The optimized program allows the more efficient and extended analysis of the generation, propagation and long time behavior of nonlinear water waves.

ACKNOWLEDGEMENTS

These investigations were supported by the German Society of Sciences (DFG) under grant No. Ma 358/48-1.

REFERENCES

[1] Coleman, Th.F. and Van Loan, Ch.:
Handbook for Matrix Computations, SIAM, Philadelphia, 1988.

[2] Rienecker, M.M. and Fenton, J.D.:
A Fourier approximation for steady water waves, J. Fluid Mech., Vol. 104, pp. 119–137, 1981.

[3] Gentzsch, W.:
Vectorization of Computer Programs with Applications to Computational Fluid Dynamics, Notes on Numerical Fluid Mechanics, Vol. 8, Vieweg, 1984.

[4] Longuet-Higgins, M.S., Cokelet, E.D.:
The deformation of steep surface waves on water. *A numerical method for computation*, Proc. Roy. Soc. London, A 350, pp. 1–26, 1976.

[5] Romate, J.E.:
Accuracy and Efficiency of a Panel Method for Free Surface Flow Problems in Three Dimensions, in: Boundary Elements IX (Eds. Brebbia, C.A., Wendland, W., Kuhn, G.), Vol. 1, pp. 229–240, Springer-Verlag, Berlin/.., 1987.

[6] Schendel, U.:
INTRODUCTION TO NUMERICAL METHODS FOR PARALLEL COMPUTERS, in: Ellis Horwood Series MATHEMATICS AND APPLICATIONS (Ed. Bell, G.M.), Ellis Horwood ltd., 1984.

[7] Wendland, W.:
Bemerkungen zu Randelementmethoden und ihren mathematischen und numerischen Aspekten, in: Mitteilungen der GAMM, Heft 2, pp. 3–27, 1986 (in german).

[8] Zucchini, A. and Mukherjee, S.:
Vectorial and Parallel Processing in Stress Analysis with the Boundary Element Method, Int. Journ. for Num. Math. in Eng., Vol. 31, pp. 307–317, 1991.

MODAL ANALYSIS OF SOLAR ARRAYS USING BOUNDARY INTEGRAL EQUATIONS

J.J. Heijstek and H. Schippers
NLR, National Aerospace Laboratory
Anthony Fokkerweg 2, 1059 CM Amsterdam, The Netherlands

ABSTRACT

In this paper the acoustic effects of the surrounding air around a harmonically vibrating solar array in a testroom are estimated. The dynamics of the solar array is modelled by a differential equation for a harmonically vibrating plate and the sound pressure of the vibrating air is modelled by a boundary integral ansatz. The contribution of the sound pressure on the solar array results in a perturbed eigenvalue problem which is approximately solved by perturbation analysis up to first order with respect to the ratio of the density of the air and the density of the solar array. This analysis involves the solution of a hypersingular integral equation which is numerically solved by a boundary element method.

INTRODUCTION

Modern spacecraft are equipped with large light-weight solar arrays, which have a strong influence on the dynamical behaviour of the spacecraft. Therefore it is a prerequisite to analyse the dynamical behaviour of the solar arrays and their interaction with the main structure.

On earth the dynamical behaviour of the solar arrays is investigated by ground vibration tests. The results of these tests are verified by performing modal analysis using a finite element model in vacuum, i.e. the effects of the surrounding air in the test environment are neglected. When the results of the ground vibration tests and the modal analysis are compared deviations of more than 10 % are observed with regard to the natural frequencies. In the present report the acoustic effects of the surrounding air around a single harmonically vibrating solar array are estimated. The mass and stiffness distribution of the solar array are such that the vibrating air behaves like added mass. It appears that the natural frequencies decrease if the air is taken into account. As a consequence, the frequencies show a better agreement with those of the ground vibration tests.

In this report the dynamics of the solar array is modelled by a differential equation for a harmonically vibrating plate and the sound pressure of the vibrating air is modelled by a boundary integral ansatz. The vibrating plate and the vibrating air are coupled by the acoustic equation. The coupling results in a differential boundary integral equation. When the differential equation is approximated by finite elements and the integral equation by boundary elements a coupled FE-BE model is obtained for the dynamics of the solar array in air.

MATHEMATICAL MODELLING

In this section we assume that the solar array can be modelled by a thin rectangular isotropic plate. For small displacements the governing differential equation for a harmonically vibrating plate is given by

$$\frac{D}{\rho}\,\Delta^2 w - \lambda^2 w = \frac{\mu}{\rho}\,, \tag{1}$$

where w is the displacement in normal direction, D the bending stiffness of the plate and ρ its density. The right-hand side of (1) represents the force acting on the plate caused by the jump in the acoustic pressure,

$$\mu = p^+ - p^-\,, \tag{2}$$

where p^+ (p^-) denotes the pressure on the upper (lower) part of the plate. When the plate vibrates in vacuum $\mu = 0$. Then (1) corresponds with a classical eigenvalue problem. The solution of this problem yields the vibration modes and the eigenfrequencies of the plate in vacuum. However, when the plates vibrates in air the pressure jump μ does not vanish in general. Below an expression for μ in terms of w is derived, by which (1) changes in a perturbed eigenvalue problem.

For the modelling of the acoustic pressure surrounding the plate we distinguish the incompressible case governed by

$$\Delta p = 0 \tag{3}$$

and the compressible case governed by

$$\Delta p + \left(\frac{\lambda}{c}\right)^2 p = 0\,, \tag{4}$$

where c the speed of sound. On the plate the acoustic pressure has to satisfy the so-called acoustic coupling equation

$$\frac{\partial p}{\partial n} = \rho_a\, \lambda^2\, w\,, \tag{5}$$

where n is the normal at the upperside and ρ_a the density of air. In the remainder of this paper the normal n is associated with the y-axis as is shown in figure 1. Far away from the plate it is required that p satisfies the Sommerfield radiation condition.

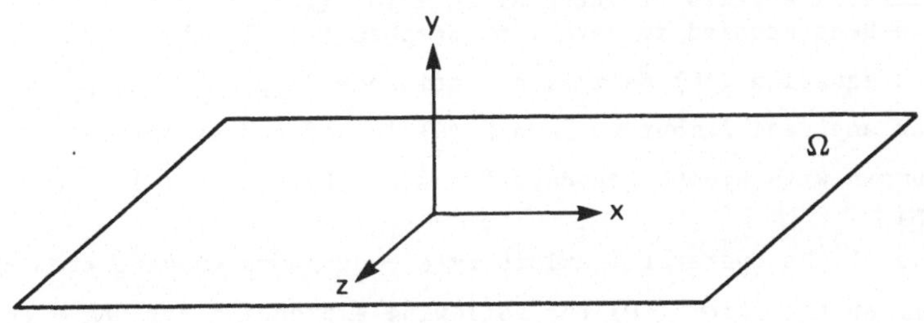

Fig. 1 Solar array with coordinate system

Note that the boundary value problem for p is related to the 3D exterior screen problems which have been studied by Stephan [1]. Here the plate corresponds with a screen, a surface in R^3. The above problem is transformed into a boundary integral equation by representing the acoustic pressure in the form of a double layer potential

$$p(\vec{r}) = \int_\Omega \frac{\partial G}{\partial y'}(\vec{r},\vec{r}')\ \mu(\vec{r}')\ d\Omega(\vec{r}')\ , \quad \vec{r} \in R^3/\Omega\ , \tag{6}$$

where Ω is the domain of the plate and G represents the Green function, which is for the incompressible case given by

$$G(\vec{r},\vec{r}') = \frac{-1}{4\pi|\vec{r} - \vec{r}'|}\ , \tag{7}$$

and for the compressible case

$$G(\vec{r},\vec{r}') = \frac{-1}{4\pi|\vec{r} - \vec{r}'|}\ e^{ik|\vec{r} - \vec{r}'|}\ ,$$

with k the acoustic wave number

$$k = \frac{\lambda}{c}\ .$$

From classical results of potential theory it is well known that

$$\mu(\vec{r}) = p^+(\vec{r}) - p^-(\vec{r})\ , \quad \vec{r} \in \Omega\ . \tag{9}$$

The application of boundary condition (5) to the double layer potential formula (6) yields the hypersingular integral equation

$$\oint_\Omega \frac{\partial^2 G}{\partial y \partial y'}\ \mu(\vec{r}')\ d\Omega(\vec{r}') = \rho_a\ \lambda^2\ w(\vec{r})\ , \quad \vec{r} \in \Omega\ , \tag{10}$$

which is written in operator notation as

$$T\mu = \rho_a\ \lambda^2\ w\ . \tag{11}$$

In (10) the integral $\oint$ is defined in the sense of Hadamard.

The mathematical aspects of integral equation (10), as defined on screens in R^3, have been studied in detail by Stephan [1]. In this paper it was proven that equation (10) defines a continuous, mapping from $\tilde{H}^s(\Omega)$ onto $H^{s-1}(\Omega)$ for any real number s. Here $\tilde{H}^s(\Omega)$ is defined as in [1]: if G is a bounded domain with smooth boundary Γ and $\Omega \subset \Gamma$, then $\tilde{H}^s(\Omega) = \{u \in H^s(\Gamma),\ \text{supp}\ u \subset \bar{\Omega}\}$.

The inverse of the operator T exists as a continuous mapping from $H^{-\frac{1}{2}}(\Omega)$ onto $\tilde{H}^{\frac{1}{2}}(\Omega)$, so that from (10) the following expression for the pressure jump is obtained

$$\mu = \rho_a \lambda^2 T^{-1} w . \tag{11}$$

Substituting (11) into (1) one obtains

$$\frac{D}{\rho} \Delta^2 w - \lambda^2 w - \varepsilon \lambda^2 T^{-1} w = 0 , \tag{12}$$

with

$$\varepsilon = \rho_a/\rho .$$

For metallic plates ε is very small and the last term in (12) may be neglected in the modal analysis. But, for solar arrays, ε is in the magnitude of 0.01 and it will be indicated in this paper that the vibrating surrounding air may have some effects on the values of the eigenfrequencies.

Equation (12) defines a compactly perturbed eigenvalue problem. Moreover, when the acoustic pressure is modelled in a comprestible medium by the Helmholtz equation the problem depends in a nonlinear way on λ, due to the occurrence of λ in the Green function via $k = \lambda/c$.

In order to obtain an expansion with respect to ε of the eigenvalues of the perturbed eigenvalue problem (12) we shall also use a variational formulation of (12), where free boundary conditions are imposed on the edge of the plate. Let the bilinear form E, associated with the first variation of the strain energy, be given by

$$E(v,w) = \frac{D}{\rho} \int_\Omega \{\sigma\, \Delta v\, \Delta w + (1-\sigma)(v_{xx}\, w_{xx} + 2\, v_{xz}\, w_{xz} + v_{zz}\, w_{zz})\}\, d\Omega$$

and let the bilinear form, related to the work due to the acoustic pressure jump, be given by

$$A(v,w) = \int_\Omega v\, T^{-1}\, w\, d\Omega . \tag{13}$$

Then, the variational formulation of (12) can be written as:
Find non-trivial $w \in H^2(\Omega)$ and $\lambda \in R^1$ such that

$$a_\varepsilon(v,w) := E\,(v,w) - \lambda^2(v,w) - \varepsilon\lambda^2 A(v,w) = 0 , \tag{14}$$

for all $v \in H^2(\Omega)$.

Next we will indicate an expansion for the perturbed eigenvalues and eigenfunctions.
Let w^o and λ^o represent a vibration mode and an eigenvalue of the problem in vacuum, i.e. w^o and λ^o satisfy

$$a_o(v,w^o) = E(v,w^o) - (\lambda^o)^2(v,w^o) = 0 \quad \text{for all } v \in H^2(\Omega) . \tag{15}$$

From standard perturbation theory (see [2, section VII.6.2]) it follows that the corresponding perturbed eigenvalues and eigenmodes of the vibrating plate in air are analytic functions of ε; the perturbed eigenvalues have the same algebraic and geometric multiplicity as the corresponding eigenvalues of the problem in vacuum. Thus, the following expansions hold:

$$\mu := \lambda^2 = \mu_o + \varepsilon\mu_1 + \varepsilon^2\mu_2 + \varepsilon^3\mu_3 + \ldots \tag{16}$$

$$w = w_o + \varepsilon\, w_1 + \varepsilon^2 w_2 + \varepsilon^3 w_3 + \ldots \tag{17}$$

for all small values of $|\varepsilon|$, where the coefficients μ_j, w_j, $j = 0,1,2,\ldots$ do not depend on ε.

Substituting the expansions (16), (17) in (12) (and (14)), and using the fact that (12) (and (14)) hold for all (sufficiently small) values of $|\varepsilon|$, one obtains

$$\mu_o = (\lambda^o)^2 \quad , \quad w_o = w^o \; ,$$

$$\mu_1 = -\,(\lambda^o)^2\, A(w^o, w^o)\,\big|_{\varepsilon=o} \,/(w^o,w^o) \; , \tag{18}$$

$$\mu_2 = (\mu_1/\lambda^o)^2 - (\lambda^o)^2 \,\{\, A(w_1,w^o) + S(w^o,w^o) \,\} \,/\, (w^o,w^o) \; , \tag{19}$$

with

$$S = \frac{\partial}{\partial\varepsilon} T^{-1} \,\Big|_{\varepsilon=o} \; ,$$

and with w_1 the solution of the problem

$$E\,(v,w_1) - (\lambda^o)^2(v,w_1) = \mu_1(v,w^o) - (\lambda^o)^2\, A(v,w^o)\big|_{\varepsilon=o} \; , \tag{20}$$

for all $v \in H^2(\Omega)$ and

$$(w_1,w^o) = 0 \; .$$

For an incompressible medium S=0 in (19) and it follows from results of Stephan [1] that the operator T is positive definite on $\tilde{H}^{\frac{1}{2}}(\Omega)$, i.e. there exists C>0 such that

$$(T\,v,v)_o \geq C\,||v||^2_{\frac{1}{2}} \quad , \quad \text{for all } v \in \tilde{H}^{\frac{1}{2}}(\Omega) \; . \tag{21}$$

Since T is bijective it follows that

$$A(w,w) \geq C\,||T^{-1}w||^2_{\frac{1}{2}} > 0 \; , \tag{22}$$

for all $w \in H^{-\frac{1}{2}}(\Omega)$, $w \neq 0$.

As a consequence we observe that $\mu_1 < 0$, i.e. the magnitudes of the eigenfrequencies of the vibrating plate in air are less than the corresponding eigenfrequencies in vacuum. For a compressible medium (21) cannot hold since in (19) (Sw^o, w^o) has nonzero imaginary part.

COUPLING FINITE ELEMENT MODEL IN VACUUM WITH BOUNDARY ELEMENTS

The vibration of a solar array in air can be represented by the general bilinear form

$$a_\varepsilon(v,w;\lambda) = E(v,w) - \lambda^2(v,w) - \varepsilon\,\lambda^2 A(v,w) \quad , \tag{23}$$

with E(v,w) the first variation of the strain energy, which is assumed to be positive definite on some Sobolev space X. Let X_h be a finite element subspace of X. Then the eigenvalue problem in vacuum reads:

<u>Find</u> $(W_h, \lambda_h) \in X_h \times R^1$ such that

$$a_o(V_h, W_h; \lambda_h) = 0 \ , \qquad \text{for all } V_h \in X_h \ . \tag{24}$$

The eigenvalues of this problem are real because of the positive definiteness of E. Let $\lambda_1 \le \lambda_2 \le \lambda_3 \ldots.$ be the eigenvalues of the problem in vacuum and let $W_1, W_2, W_3, \ldots.$ be the corresponding eigenfunctions.

Introduce a subspace $\tilde{X}_N$ of eigenfunctions by

$$\tilde{X}_N = \text{span}\,\{\, W_1, W_2, \ldots\ldots, W_N \,\}.$$

Then, the approximation of the eigenvalue problem in air reads:

<u>Find</u> $(W,\lambda) \in \tilde{X}_N \times R^1$ such that

$$a_\varepsilon\,(v,W;\lambda) = 0 \ , \qquad \text{for all } V \in \tilde{X}_N \ . \tag{25}$$

From the asymptotic analysis of the previous section we have the first order expansion

$$(\lambda^2)_{air} = (1-\varepsilon\alpha)\ (\lambda^2)_{vacuum} + O(\varepsilon^2)(\varepsilon\to 0) \tag{26}$$

with

$$\alpha = \frac{A(W,W)}{(W,W)} \ , \ W \in \tilde{X}_N \ . \tag{27}$$

Here W denotes a finite element mode of the vibrating plate in vacuum. In order to get a qualitative insight in the magnitude of α we have to determine the numerator of α. The bilinear form A, which has been defined in (13), involves the determination of

$$\sigma = T^{-1}\, W,$$

which follows from solving the boundary integral equation

$$T\sigma = W. \tag{28}$$

This boundary integral equation is solved by boundary elements. Introduce a boundary element space $Y_h \subset \tilde{H}^{\frac{1}{2}}(\Omega)$ and find $\sigma_h \in Y_h$ such that

$$(T\sigma_h, \mu_h) = (W, \mu_h) \ , \quad \text{for all } \mu_h \in Y_h. \tag{29}$$

In this way the estimate of the acoustic effects can be treated as postprocessing by boundary element calculations, after the modal analysis of the finite element model in vacuum has been performed.

NUMERICAL APPROXIMATION

The hypersingular integral equation (10) is regularized by integration of parts (see e.g. [3]) using the boundary condition that the jump in the sound pressure vanishes along the edge of Ω,

$$\mu(\vec{r}) = 0 \ , \quad \vec{r} \in \partial\Omega \ . \tag{30}$$

Then the bilinear form

$$(T\mu, \sigma) = \iint\limits_{\Omega\Omega} \frac{\partial G}{\partial y \partial y'} \mu(\vec{r}\,') \ \sigma(\vec{r}\,') \ d\Omega(\vec{r}\,') \ d\Omega(\vec{r}) \ ,$$

which is used in (29), can be rewritten as

$$(T\mu, \sigma) = k^2 \iint\limits_{\Omega\Omega} G(\vec{r}, \vec{r}\,') \ \mu(\vec{r}\,') \ \sigma(\vec{r}) \ d\Omega(\vec{r}\,') \ d\Omega(\vec{r})$$

$$- \iint\limits_{\Omega\Omega} G(\vec{r}, \vec{r}\,') \left(\frac{\partial\mu}{\partial x'} \frac{\partial\sigma}{\partial x} + \frac{\partial\mu}{\partial z'} \frac{\partial\sigma}{\partial z}\right) d\Omega(\vec{r}\,') \ d\Omega(\vec{r}) \ . \tag{31}$$

When the air is assumed to be incompressible the first term in (31) vanishes. Observe that the regularization leads to weakly singular integrals which are of benifit in evaluating the influence coefficients of the systemmatrix corresponding to (29). Let the boundary element space be given by piecewise linear basisfunctions on triangular elements Ω_i, $i=1(1)N$, covering the region Ω as shown in figure 2.

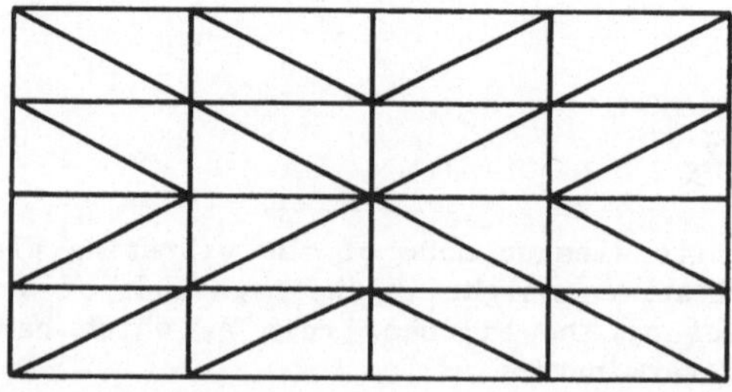

Fig. 2 Triangular grid on solar array

Define on Ω_i the following approximation

$$\sigma_h^i(\vec{r}) = \sum_{\ell=1}^{3} \sigma_\ell^i \, \psi_\ell^i(\vec{r}) \quad , \tag{32}$$

where $\psi_\ell^i(\vec{r})$, $\ell = 1(1)3$, are the linear element shape functions with support in Ω_i. Furthermore, σ_1^i , σ_2^i and σ_3^i are the values of $\sigma(\vec{r})$ at the vertices of the triangular element Ω_i. Then, the bilinear form $(T\mu,\sigma)$ can be approximated by

$$(T\mu,\sigma) = \sum_{i=1}^{N} \sum_{\ell=1}^{3} \sum_{j=1}^{N} \sum_{m=1}^{3} K_{\ell,m}^{i,j} \, \sigma_\ell^i \, \mu_m^j \, , \tag{33}$$

where the influence coefficients are given by

$$K_{\ell,m}^{i,j} = k^2 \int_{\Omega_i}\int_{\Omega_j} G(\vec{r},\vec{r}\,') \, \psi_k^i(\vec{r}) \, \psi_\ell^j(\vec{r}\,') \, d\Omega(\vec{r}) \, d\Omega(\vec{r}\,')$$

$$- \int_{\Omega_i}\int_{\Omega_j} G(\vec{r},\vec{r}\,') \left\{ \frac{\partial\psi_\ell^i}{\partial x} \frac{\partial\psi_m^j}{\partial x'} + \frac{\partial\psi_\ell^i}{\partial y} \frac{\partial\psi_m^j}{\partial y'} \right\} d\Omega(\vec{r}) d\Omega(\vec{r}\,') \ . \tag{34}$$

For the evalution of the influence coefficients we distinguish regular integrals ($i \neq j$) and weakly singular integrals ($i=j$). The regular integrals are approximated by one-point Gauss quadrature formulas on the midpoints of the elements (See e.g. [4]). The weakly singular integrals are of the following form

$$I(\vec{r}) = \int_{\Omega_i} \frac{1}{4\pi|\vec{r}-\vec{r}\,'|} \, f(\vec{r}\,') \, d\Omega(\vec{r}\,') \ . \tag{35}$$

Introduce a local coordinate system (s,t) on Ω_i and let $\vec{r} \in \Omega_i$ be given by $\vec{r} = \vec{r}(s,t)$. When we denote $\vec{r}\,' = \vec{r}(\xi,\eta)$, then the distance between $\vec{r}$ and $\vec{r}\,'$ becomes

$$|\vec{r}-\vec{r}\,'| = \{((x_1-x_3)(\xi-s) + (x_2-x_3)(\eta-t))^2$$

$$+((y_1-y_3)(\xi-s) + (y_2-y_3)(\eta-t))^2 \}^{\frac{1}{2}} =: \rho(s,t;\xi,\eta).$$

When we substitute this expression into (35) we obtain

$$I(\vec{r}(s,t)) = \iint_{\xi\eta} \frac{1}{4\pi\rho(s,t;\xi,\eta)} \, f(\vec{r}(\xi,\eta)) \, J(\xi,\eta) \, d\xi \, d\eta, \tag{36}$$

where $d\Omega(\vec{r}') = J\, d\xi\, d\eta$. Let us now introduce polar coordinates (a,φ) centered at (s,t). Then

$$\rho(s,t;\xi,\eta) = a\,\{((x_1-x_3)\cos\varphi + (x_2-x_3)\sin\varphi)^2$$

$$+((y_1-y_3)\cos\varphi + (y_2-y_3)\sin\varphi)^2\}^{\frac{1}{2}} =: aR(\varphi)$$

and $d\xi\, d\eta = a\, da\, d\varphi$. Substituting these expressions into (36) we obtain

$$I(\vec{r}(s,t)) = \int_0^{2\pi}\int_0^{\delta(\varphi)} \frac{1}{4\pi R(\varphi)} f(a,\varphi)\, J(a,\varphi)\, da\, d\varphi , \tag{37}$$

where $\delta(\varphi)$ is the representation in polar coordinates of the edges of Ω_i. Now, we observe that the integrand in (37) is regular, so that we can compute the double integral in (37) by standard quadrature rules.

Next, we are going to discuss another numerical approximation by defining the boundary element space by piecewise bilinear functions on rectangular elements as shown in figure 3. The gradient of σ in (31) is approximated by piecewise constant functions, i.e.,

$$\sigma_x := (\sigma_{i+1,j} - \sigma_{i,j} + \sigma_{i+1,j+1} - \sigma_{i,j+1}) \,/\, 2\,\Delta x , \tag{38}$$

where $\Delta x = x_{i+1,j} - x_{i,j}$. An analogous approximation holds for σ_y. When the regular integrals are again evaluated by one-point Gauss quadrature and the weakly singular integrals by analytical integration, then this second approach can be written in variational formulation as follows. Let X_h be the finite dimensional space of piecewise constant functions on Ω_i and let Y_h be the finite element space of bilinear functions. For the incompressible case e.g. the variational formulation reads:

<u>Find</u> $\sigma \in Y_h$ and $\vec{t} \in X_h \times X_h$ such that

$$-\iint_{\Omega\Omega} G(\vec{r},\vec{r}')(t_1\mu_{x'} + t_2\mu_{y'})\, d\Omega(\vec{r})\, d\Omega(\vec{r}') = (W,\mu), \text{ for all } \mu \in Y_h, \tag{39}$$

and

$$\int_\Omega (\vec{t} - \nabla\sigma).\vec{s}\, d\Omega(\vec{r}) = 0, \text{ for all } \vec{s} \in X_h \times X_h . \tag{40}$$

When the latter integral is evaluated by one-point Gauss quadrature at the midpoint of Ω_i, the approximation of (38) is obtained. The variational formulation (39)-(40) can be considered as a mixed finite element approximation for the hypersingular boundary integral equation (28). The stability of this approach will be analyzed in a forthcoming paper.

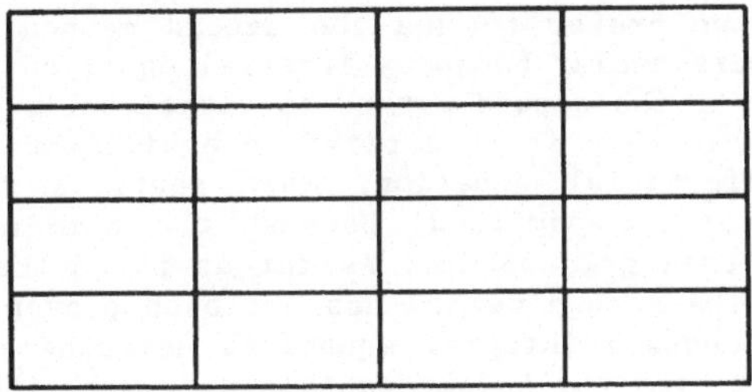

Fig. 3 Rectangular grid on solar array

NUMERICAL RESULTS

The consistency of the boundary element methods of the previous section is verified by computing $(\mu,T\sigma)$, where μ and σ are given by

$$\mu(x,y) = \sigma(x,y) = \cos(\frac{\pi x}{a}) \cos(\frac{\pi y}{b}), \tag{41}$$

with a=2, b=1 and the Green function given by (7). Here, it is assumed that the domain Ω is defined by the rectangle $[-a/2,a/2] \times [-b/2,b/2]$. The grids on Ω are obtained by dividing the domain into uniform rectangular elements with $\Delta x=a/N$ and $\Delta y=b/N$. The triangular grid on Ω is obtained by dividing the rectangular elements once again as indicated in Figure 2. Observe that the number of nodes in both grids are equal, so that the number of degrees of freedom in the two boundary element methods of the previous section are also the same.

In table 1 we give the numerical results for N=8, 16 and 32. The exact value of $(\mu,T\sigma)$ is unknown, but is was computed to be 0.733 using fine grids and extrapolation techniques. When we compare the results of table 1 with this value, we conclude that the second method (using numerical differentiation formulas for $\nabla\sigma$) seems to converge quadratically, while the first method (using piecewise linear shape functions on triangular elements) converges linearly.

Table 1 Approximate values of $(\mu,T\sigma)$
method 1: piecewise linear functions on triangular grid
method 2: bilinear test funtions and numerical differentiation formulas on rectangular grid
N : number of segments in one direction

N	method 1	method 2
8	.607	.699
16	.673	.724
32	.705	.731

Inspecting table 1 we prefer to use the second method for the numerical solution of the hypersingular boundary integral equation (28). However, the numerical stability of the second method is questionable because the method appears to be unstable when it is applied to a bilinear form corresponding with Laplace's differential equation. Then spurious modes (oscillatory "sawtooth" solutions) are admitted, because the numerical scheme is decoupled in odd and even grid points. As far as the authors know the numerical stability of the second method has not been proven theoretically for the hypersingular boundary integral equation. Nevertheless, it is applied here to equation (28) for the incompressible case with the vibration mode W given by the right-hand side of (40), i.e. W corresponds with the (1,1)-mode of a simply supported plate in vacuum. The integral equation (28) was solved using 16 segments in both x- and y-direction, which results in 289 degrees of freedom. For this subdivision the system of equations corresponding to (39)-(40) could be solved without any problem on the NLR Cyber 962 computer. Hence, possible numerical instability of the second method was not established so far by computations. The value of α in (26)-(27) was computed to be $\alpha = 0.69$.

Substituting this value into (26) and taking $\varepsilon=0.02$ we observe that the natural frequency of a simply supported harmonically vibrating plate decreases with 0.7 % for the (1,1)-mode when the acoustic pressure of the air is modelled by Laplace's equation. This effect is rather small in comparison with the 10 % deviation mentioned in the introduction. The latter deviation, however, is observed when two stowed solar arrays harmonically vibrate close to each other out of phase. The modal analysis of such stowed solar arrays as well as the effects of the compressibility of the air (i.e. modelling the acoustic pressure by (4) instead of (3)) will be investigated in future research.

CONCLUDING REMARKS

A procedure for estimating the acoustic effects of the surrounding air around a harmonically vibrating plate has been presented. Using standard perturbation theory a first order aprroximation has been derived for the deviation of the natural frequencies of the model in air and the ones of the model in vacuum. The calculation of the acoustic effects is performed as postprocessing on the results of the modal analysis in vacuum. The procedure involves the numerical solution of a hypersingular boundary integral equation of which the right-hand side is prescribed by some eigenmode of the modal analysis in vacuum. For the numerical solution two boundary element methods were considered. The first one uses piecewise linear functions on a triangular grid, while in the second one the gradient of the pressure jump is approximated by numerical differentation formulas on a rectangular grid. The numerical stability of the second method is questionable; it will be investigated in future research. Preliminary numerical results reveal that the second method yields more accurate results than the first one.

REFERENCES

[1] STEPHAN, E.P.: "Boundary integral equations for screens in R^3", Integral Equations and Operator Theory, 10 (1987), pp. 236-257.

[2] KATO, T.: "Perturbation theory for linear operators; second corrected printing of the second edition", Springer Verlag, Berlin, Heidelberg, New York, Tokyo, 1984.

[3] NEDELEC, J.C.: "Approximation par potential de double couche du problème de Neumann extérieur", C.R. Acad. Sci. Paris, Sér. A, 1977, 286, pp. 616-619.

[4] BECKER, E.B., CAREY, G.F., ODEN, J.T.: "Finite elements, An introduction, Volume 1", Prentice-Hall, Inc., Englewood Cliffs, New Jersey, 1981.

A Panel Method Using Numerical Integration

Gerhard Jensen
Hamburgische Schiffbau-Versuchsanstalt GmbH

1. Introduction

Panel methods are used to determine the singularity distribution (sources in most cases) on a body surfaces so that the Neumann condition - no flow across the body boundary -

$$\vec{n}\nabla\Phi = v_n \tag{1}$$

is fulfilled on the body surface S. Φ is the potential due to the singularity distribution, $\vec{n}$ is the normal vektor on the surface, v_n is the normal component of the incoming flow (e.g. $-Un_1$ for a body in a parallel stream against the x-axis). The Neumann condition is required at discrete control points in most cases to determine the intensity of the singularity distribution. In case of a smooth surface and a smooth singularity distribution it can be assumed, that the Neumann condition is approximatly fulfilled at the other points on the surface as well.

The well known method of Hess&Smith [1] uses a source distribution of constant strength within plane panels. The control points are at the center of the panel and the body normal is approximated by the panel normal.

Most panel methods approximate the distribution of the source strength within each panel by a low order polynomial, in the case of Hess&Smith a constant. Thus the integral of the Greensfunction can be determined piece-wise analytically. The numerical evalution of these analytic expressions requires the evaluation of transcendent function to a high degree of accuracy, which requires a significant amount of computer time. Therefore the idea arose to replace the analytic integration by a simple numerical integration. One of the problems that has to be overcome to this end is, that the integrand does not remain finite as the source point approaches the control point. In the following methods are shown to overcome the problem for both, the 2-dimensional and the 3-dimensional flow.

2. Problem

The potential due to a source distribution on the body surface is:

$$\phi(\vec{q}) = \int_S M(\vec{p})\, G(\vec{p},\vec{q})\, dS_p\,. \tag{2}$$

$\phi(\vec{q})$ is the potential at a point $\vec{q}$ induced by a source distribution M on the body surface S; $G(\vec{p},\vec{q})$ is the potential of the unit source at $\vec{p}$: $G = \log|\vec{p}-\vec{q}|/2\pi$ for 2-d flow and $G = -(4\pi|\vec{p}-\vec{q}|)^{-1}$ for 3-d flow. The normal velocity on the body surface S is

$$v_n(\vec{q}) = \vec{n}(\vec{q})\,\nabla_q\phi(\vec{q}) = \int_S M(\vec{p})\,\vec{n}(\vec{q})\,\nabla_q G(\vec{p},\vec{q})\, dS_p - \frac{1}{2}M(\vec{q}). \tag{3}$$

If the normal velocity is prescribed by the boundary condition, the important part of the solution is the velocity in the tangential direction $\vec{t}$ (The velocity components in two different tangentialial directions have to be determined in the 3-d case):

$$v_t = \vec{t}(\vec{q})\,\nabla_q\phi(\vec{q}) = \oint_S M(\vec{p})\,\vec{t}(\vec{q})\,\nabla_q G(\vec{p},\vec{q})\, dS_p. \tag{4}$$

In case of 2-dimensional flow on finite curvature of the body surface the integrand in (3) approaches a constant as $\vec{p} \to \vec{q}$ see Jensen [2]; (4) is singular. For a 3-dimensional flow the integrand in (3) is singular for $\vec{p} \to \vec{q}$ for a curved surface; inside a plane panel it tends to zero, but the integrand in (4) is unbounded even within a plane panel with constant source strength.

Due to this behaviour there is no difficulty to evaluate (3) numerically, although it might be improved, but (4) can not be used directly.

3. Modifications to the equations

In the first step we try to improve the properties of the integrand in equation (3). To this end the following properties are used:
$G(\vec{p},\vec{q})$ could also be interpreted as potential at $\vec{p}$ due to a source at $\vec{q}$:

$$\nabla_q G(\vec{p},\vec{q}) = -\nabla_p G(\vec{p},\vec{q})\,. \tag{5}$$

The total flux across the boundary of a closed domain due to a source on its smooth surface is equal to half the productivity (strength) of the source:

$$\int_S \vec{n}(\vec{p})\,\nabla_q G(\vec{p},\vec{q})\,dS_p = -\int_S \vec{n}(\vec{p})\,\nabla_p G(\vec{p},\vec{q})\,dS_p = \frac{1}{2}\,. \tag{6}$$

This expression times $M(\vec{q})$ can be added to (3):

$$v_n(\vec{q}) = \int_S [M(\vec{p})\,\vec{n}(\vec{q}) + M(\vec{q})\,\vec{n}(\vec{p})]\ \nabla_q G(\vec{p},\vec{q})\,dS_p - M(\vec{q}). \tag{7}$$

It can be shown, that the integrand vanishes a for smooth contour and a smooth source distribution in the 2-dimensional case and therefore has excellent properties for numerical integration.

In the 3-dimensional case the integrand approaches a constant that depends on the direction of the approach. Only within a plane panel it goes to zero within the panel, but that is no improvement over the direct application of (3).

3.1 Tangential velocity for the 2-dimensional case

We use the fact, that the circulaiton in a flow induced by sources is zero and (5):

$$\int_S \vec{t}(\vec{p})\,\nabla_q G(\vec{p},\vec{q})\,dS_p = -\int_S \vec{t}(\vec{p})\,\nabla_p G(\vec{p},\vec{q})\,dS_p = 0\,. \tag{8}$$

This transforms (4) into

$$v_t = \oint_S \left[M(\vec{p})\,\vec{t}(\vec{q}) - M(\vec{q})\,\vec{t}(\vec{p})\right]\ \nabla_q G(\vec{p},\vec{q})\,dS_p. \tag{9}$$

The integrand now remains finite as $\vec{p} \to \vec{q}$. If panels with constant source strength are used the integrand vanishes within the panel, thus numerical integration can be used to evaluate (9)

3.2 Tangential velocity for 3-dimensional flow

A source distribution of constant strength on the surface **S** of a sphere does not induce any tangential velocity on **S**:

$$\int_{\mathbf{S}} \vec{t}(\vec{q})\nabla_q G(\vec{k},\vec{q})\,d\mathbf{S}_k = 0 \ \text{ for } \vec{q} \text{ and } \vec{k} \text{ on } \mathbf{S}. \tag{10}$$

The sphere is placed touching the body tangentially at the point $\vec{q}$. If the centre of the sphere is within the body, there exists a projection $\vec{k} = P(\vec{p})$ of every point $\vec{p}$ on the body surface to a unique point $\vec{k}$ on the sphere surface, the projection being defined by a straight line passing through $\vec{k}$, $\vec{p}$ and the sphere's centre. The projection of all body points will cover the whole sphere surface at least once. Surface elements $d\mathbf{S}_k$ on the sphere are projected onto surface elements dS_p on the body. Let r be the area ratio of the surface elements: $d\mathbf{S}_k = r dS_p$. The sign of r is defined by the sign of the scalar product of the corresponding normal vectors pointing into the body or sphere, respectively. With these definitions, (10) can be transformed into an integral over the body surface:

$$\oint_S \vec{t}(\vec{q})\, \nabla_q G(P(\vec{p}), \vec{q})\, r\, dS_p = 0. \tag{11}$$

This expression is multiplied by $M(\vec{q})$ and subtracted from (4):

$$v_t = \vec{t}(\vec{q})\, \nabla_q \phi(\vec{q}) = \oint_S \Big[M(\vec{p})\, \vec{t}(\vec{q})\, \nabla_q G(\vec{p}, \vec{q}) - $$

$$M(\vec{q})\, \vec{t}(\vec{q})\, \nabla_q G(P(\vec{p}), \vec{q})\, r \Big]\, dS_p. \tag{12}$$

If $\vec{p}$ approaches $\vec{q}$ the integrand is still singular in general, but within a panel of constant source strength it approaches zero. Details can be found in [2].

4. Discretization and some results

In the following only the more relevant 3-dimensional case is described. To evaluate (3) and (12) numerically the closed surface of the body is discretized into N panels. For each of them area f_i, midpoint $\vec{x}_i$, unit normal $\vec{n}_i$ and two approximately orthogonal tangent vectors $\vec{s}_i$ and $\vec{t}_i$ are determined, and the radius of the tangential sphere is chosen. The radius turned out to have negligible influence on the results within wide limits. At each point the normal velocity v_n is prescribed (e.g. $v_n = U n_1$).

The panel midpoints $\vec{x}_i$ are used both as control (collocation) points where the boundary condition is fulfilled, and as integration points for the numerical integration over the body surface, which is performed simply by adding the products of integrand times panel area. Thus (3) gives a system of linear equations for the unknown source strengths, which can be solved using simple Gauss-Seidel iteration. Afterwards (12) is used to determine the unknown velocities.

Test computations for a sphere in a parallel flow show that:

- the error decreases proportional to the grid spacing (1st order method)
- the result only very weakly depends on the chosen radii for the tangential spheres
- the error has approximately the same magnitude as in the widely used method of Hess&Smith although the computiong time is shorter
- the error is significantly larger than with Webster's Method (submerged triangular panels with linearily varying source strengths and and collocation points at the corners).

A sphere however is not really a practical problem. In Naval Hydrodynamics we are often interested in the flow around bodies which are not as smooth as spheres. They may have sharp corners at least on a plane of symmetry as in the case of a ship with sharp stem. We know that two of the assumptions of ordinary panel methods: e.g. smoothness of the surface and smoothness of the source strength are not fulfilled. From an engineering point of view we are however satisfied if the solution is valid on most of the hull with the exception of a small area in the vicinity of the corner. Test computations for the Wigley hull

$$Y = .8(1 - Z^2)(1 - \frac{x^2}{64}), \quad 0 \le Z \le 1, \quad -8 \le X \le 8 \tag{13}$$

in a parallel stream show convergent results for the new method as well as for the Hess&Smith method. Webster's method, which works so nicely for the sphere fails to give plausible and convergent results unless the corners are very carefully rounded. Thus this method is not practically applicable.

5. Conclusion

Methods have been shown that allow numerical integration for both 2-dimensional and 3 dimensional flow. The panel method is as accurate as Hess&Smith's method, but is computationally more efficient and more flexible because it can work with any number of corners per panel. It requires however a closed body and has certain limitations concerning the geometry. The method has proven to be a valid tool in many applications.

References

[1] HESS, J. L., SMITH, A. M. O.: "Calculation of Non-Lifting Potential Flow about Arbitrary Three-Dimensional Bodies", Douglas Aircraft Division Report No. E.S.40622, 1962.

[2] JENSEN, G.: "Berechnung der stationären Potentialströmung um ein Schiff unter Berücksichtigung der nichtlinearen Randbedingung an der freien Wasseroberfläche", Institut für Schiffbau Hamburg, Report No. 484, Juli 1988.

ON THE EXISTENCE AND EVALUATION OF THE DERIVATIVES OF THE SINGLE LAYER POTENTIAL

K.Kalik
Mathematisches Institut A, Universität Stuttgart
Pfaffenwaldring 57, D-W-7000 Stuttgart 80, Germany.

SUMMARY

Boundary element methods sometimes require the values of the derivatives of the single layer potential on the boundary in consideration. Since this derivatives are expressed by singular integrals, the regularisation of these integrals, including its calculation, is necessary. This problem is considered in this paper in the case of a two dimensional surface in $\mathbb{R}^3$.

Acknolegement: This work was supported by the Priority Research Programme "Boundary Element Methods" of the German Foundation DFG under Grant Nb. 659/18-2.

Consider the simple layer potential

$$V(x) = \iint_\Gamma \frac{\mu(y)}{|y-x|}\, d\Gamma_y, \tag{1}$$

where $\Gamma \subset \mathbb{R}^3$ is a two dimensional surface and $\mu : \Gamma \longrightarrow \mathbb{R}$ is a given function. Let s be an arbitary direction in $\mathbb{R}^3$, given by the vector

$$s = (s_1, s_2, s_3) \in \mathbb{R}^3 \;:\; s_1^2 + s_2^2 + s_3^2 = 1.$$

The aim of this paper is studying the existence and numerical calculation of the derivative

$$\frac{dV(x^0)}{ds}$$

for an arbitary point $x^0 \in \Gamma$.
Since affine coordinate mappings transform every single layer potential in a single layer potential, we may assume that the coordinate system Oy_1, Oy_2, Oy_3 is a local coordinate system which, together with the surface Γ, satisfies the following conditions:
(H.1)

i. The center of the coordinate system Oy_1, Oy_2, Oy_3 is the point x^0:

$$x^0 = (0,0,0).$$

ii. There exist $h_0 > 0$ and $\Gamma_0 \subset \Gamma$, such that

a.) $x^0 \in \Gamma^0$ and

b.) the equation pf the part Γ_0 of the surface Γ has the form:

$$y_3 = f(y_1, y_2)\,, \quad (y_1, y_2) \in K_{h_0}\,,$$

where

$$K_{h_0} := \left\{(y_1, y_2) \in \mathbb{R}^2 \,|\, y_1^2 + y_2^2 \leq h_0^2\right\}.$$

iii. if Γ_0 has a tangential plane in x^0, then the equation of this plane is $y_3 = 0$.

Let ρ and ϕ be the polar-coordinates in the plane $y_3 = 0$: $y_1 = \rho\cos\phi\,,\ y_2 = \rho\sin\phi$ and let be

$$f(\rho, \phi) := f(\rho\cos\phi, \rho\sin\phi).$$

In addition to the condition (H.1) we impose the following conditions:
(H.2)

i. $f \in C(K_{h_0})\,,$

ii. for every fixed $\phi \in [0, 2\pi]$, f as a function of ρ belongs to the class $C^1[0, h_0)$, and

iii. for every fixed $\rho \in [0, h_0)$, f as function of ϕ is continuous and piecewice continuously differentiable.

All these conditions are satisfied by the following examples:

Example 1. $f = 0$, i.e. $\Gamma_0 = K_{h_0}$.

Example 2. $f \in C^1(K_{h_0})$

Example 3. $f(y_1, y_2) = \sqrt{y_1^2 + y_2^2}$. In this case Γ_0 is a cone with vertex in x^0.

Example 4. The function f is piecwice linear and continuous. Such a function is as follows:

$$\begin{aligned}
f(y_1, y_2) &= A^{(1)}y_1 + B^{(1)}y_2 \quad \text{for} \quad y_1 \geq 0\,,\ y_2 \geq 0\,,\\
f(y_1, y_2) &= A^{(2)}y_1 + B^{(2)}y_2 \quad \text{for} \quad y_1 \leq 0\,,\ y_2 \geq 0\,,\\
f(y_1, y_2) &= A^{(3)}y_1 + B^{(3)}y_2 \quad \text{for} \quad y_1 < 0\,,\ y_2 < 0\,,\\
f(y_1, y_2) &= A^{(4)}y_1 + B^{(4)}y_2 \quad \text{for} \quad y_1 > 0\,,\ y_2 < 0\,.
\end{aligned}$$

The image of this function is a part of a polyhedron with the vertex x^0.
In the following we shall always assume, that the conditions (H.1) and (H.2) are satisfied.
In this case exists

$$f_0(\phi) := \lim_{\rho\to 0} f(\rho, \phi)$$

and

$$f(\rho, \phi) = f_0(\phi) + \rho f_1(\rho, \phi) \tag{2}$$

Be

$$V_0(x) := \iint_{\Gamma_0} \frac{\mu(y)}{|y - x|}\, d\Gamma_y\,. \tag{3}$$

The derivative $\frac{dV(x^0)}{ds}$ exists, if $\frac{dV_0(x^0)}{ds}$ exists. Therefore, we shall study the existence of the derivative of V_0 in x^0. Consequently we study the existence of the limit

$$\lim_{y^s \to x^0} \frac{V_0(y^s) - V_0(x^0)}{|y^s|} ,$$

where y^s are points of the line $\{y \in \mathbb{R}^3 \,|\, \exists \lambda \in \mathbb{R} : y = \lambda s\}$. Let r_s, θ_s and ϕ_s be the spherical coordinates of the point y^s:

$$y_1^s = r_s \sin\theta_s \cos\phi_s , \qquad y_2^s = r_s \sin\theta_s \sin\phi_s , \qquad y_3^s = r_s \cos\theta_s .$$

For arbitrary $y \in \Gamma_0$ let

$$y_1 = r \sin\theta \cos\phi , \qquad y_2 = r \sin\theta \sin\phi , \qquad y_3 = r \cos\theta .$$

The equation of Γ_0 in spherical-coordinates may be expressed in the following form:

$$\sin\theta = \frac{\rho}{\sqrt{\rho^2 + f^2(\rho, \phi)}} \tag{4}$$

The formula (2) shows, that for each $\phi \in [0, 2\pi]$

$$\lim_{\rho \to 0} \sin\theta = \frac{1}{\sqrt{1 + f_0^2(\phi)}} .$$

exists. Let θ_0 denote the angle defined by the equation

$$\sin\theta_0 = \frac{1}{\sqrt{1 + f_0^2(\phi)}} .$$

Concerning our examples we note that $\theta_0 = \frac{\pi}{2}$ in examples 1 and 2, $\theta_0 = \frac{\pi}{4}$ in example 3 and in example 4 the angle θ_0 is a continuous function of ϕ. Further note that

$$\sin\theta = \sin\theta_0 + O(\rho) . \tag{5}$$

Let γ denote the angle between the vectors y and y^s. Hence,

$$\cos\gamma = \sin\theta \sin\theta_s \cos(\phi - \phi_s) + \cos\theta \cos\theta_s .$$

Let in addition

$$\cos\gamma_0 = \sin\theta_0 \sin\theta_s \cos(\phi - \phi_s) + \cos\theta_0 \cos\theta_s .$$

We shall transform the surface integrals $V_0(y^s)$ and $V_0(x^0)$ into double integrals over K_{h_0}. For all $y \in \Gamma_0$ we denote

$$\rho := r \sin\theta .$$

Then we have $y_1 = \rho\cos\phi$ and $y_2 = \rho \sin\phi$. Further denote

$$\tilde{\mu}(y_1, y_2) := \mu\left(y_1, y_2, f(y_1, y_2)\right) \sqrt{1 + f_{y_1}^2(y_1, y_2) + f_{y_2}^2(y_1, y_2)}$$

and

$$\tilde{\mu}(\rho,\phi) := \tilde{\mu}(\rho\cos\phi, \rho\sin\phi)\,.$$

By means of the equality

$$|y-y^s| = \sqrt{r^2 - 2rr_s\cos\gamma + r_s^2} = \sqrt{\left(\frac{\rho}{\sin\theta}\right)^2 - 2\frac{\rho}{\sin\theta}r_s\cos\gamma + r_s^2}$$

we obtain

$$V(y^s) = \int_0^{2\pi} d\phi \int_0^{h_0} \frac{\rho}{\sqrt{\left(\frac{\rho}{\sin\theta}\right)^2 - 2\frac{\rho}{\sin\theta}r_s\cos\gamma + r_s^2}}\, d\rho \tag{6}$$

and

$$V(x^0) = \int_0^{2\pi} d\phi \int_0^{h_0} \tilde{\mu}(\rho,\phi)\sin\theta\, d\rho\,. \tag{6'}$$

We shall always assume that the function μ is at least continuous. This implies the existence of

$$\tilde{\mu}_0(\phi) = \lim_{\rho\to 0}\tilde{\mu}(\rho,\phi)$$

and that

$$\tilde{\mu}(\rho,\phi) = \tilde{\mu}_0(\phi) + O(\rho)\,.$$

Two remarks concerning this equality: First, $f \in C^1(K_0)$ implies $\tilde{\mu}_0(\phi) = \mu(x^0)$ for all $\phi \in [0, 2\pi]$ and second, $\mu \in C^1$ implies

$$\tilde{\mu}(\rho,\phi) = \tilde{\mu}_0(\phi) + \rho\tilde{\mu}(\rho,\phi)\,.$$

Concerning the existence of the derivative $\frac{dV_0(x^0)}{ds}$ we shall prove the following two theorems:

Theorem 1. *Let (H.1) and (H.2) be satisfied. Then for all continuous functions μ we obtain:*

i. The derivative $\frac{dV_0(x^0)}{ds}$ exists iff the limit

$$\lim_{r_s\to 0}\int_0^{2\pi} d\phi \int_{r_s}^{h_0} \tilde{\mu}(\rho,\phi)\frac{\cos\gamma\sin^2\theta}{\rho}\, d\rho$$

exists.

ii. The existence of $\frac{dV_0(x^0)}{ds}$ implies the equality

$$\begin{aligned}\frac{dV_0(x^0)}{ds} &= \lim_{r_s\to 0}\int_0^{2\pi} d\phi \int_{r_s}^{h_0} \tilde{\mu}(\rho,\phi)\frac{\cos\gamma\sin^2\theta}{\rho}\, d\rho - \\ &- \int_0^{2\pi} \tilde{\mu}(\phi)\sin^2\theta_0 \cdot \left\{1 + \cos\gamma_0 + \cos\gamma_0 \ln\frac{(1-\cos\gamma_0)\sin\theta_0}{2}\right\} d\phi\,.\end{aligned} \tag{7}$$

Theorem 2. *Let in addition to the conditions of Theorem 1 $f \in C^1(K_{h_0})$ and $\mu \in C^1$ be valid. Then we obtain:*

i. The derivative $\frac{dV_0(x^0)}{ds}$ exists and

ii.

$$\frac{dV_0(x^0)}{ds} = \lim_{r_s \to 0} \int_0^{2\pi} d\phi \int_{r_s}^{h_0} \tilde{\mu}(\rho,\phi) \frac{\cos\gamma \sin^2\theta}{\rho} \, d\rho - 2\pi\mu(x^0)\cos\theta_s \quad . \tag{8}$$

Proof of Theorem 1: First we replace ρ by t in (6) and (6′) by means of $\rho = r_s t$. Then we use the notations

$$q = \frac{t}{\sin\theta} \quad \text{and} \quad w(q) = \sqrt{q^2 - 2q\cos\gamma + 1}\,.$$

Thus we get the equality

$$\frac{V_0(y^s) - V_0(x^0)}{r_s} = \int_0^{2\pi} d\phi \int_0^{\frac{h_0}{r_s}} \tilde{\mu}(r_s t, \phi) \left[\frac{t}{w(q)} - \frac{t}{q} \right] dt \tag{9}$$

Moreover $0 < r_s < h_0$ implies $\frac{h_0}{r_s} > 1$ and hence we may rewrite (9) by

$$\begin{aligned} \frac{V_0(y^s) - V_0(y^0)}{r_s} &= \int_0^{2\pi} d\phi \int_0^1 \tilde{\mu}(r_s t, \phi) \left[\frac{t}{w(q)} - \frac{t}{q} \right] dt + \\ &\quad + \int_0^{2\pi} d\phi \int_1^{\frac{h_0}{r_s}} \tilde{\mu}(r_s t, \phi) \left[\frac{t}{w(q)} - \frac{t}{q} \right] dt \,. \end{aligned} \tag{9'}$$

For all $t > 1$ we have $q = \frac{t}{\sin\theta} > 1$ and hence $\frac{1}{q} < 1$. This inequality permits the use of the following well known representation

$$\frac{1}{w(\frac{1}{q})} = \sum_{n=0}^{\infty} P_n(\cos\gamma) \cdot \frac{1}{q^n}$$

for $t > 1$, where P_n is the Legendre polynomial of degree n. By means of these formulas we obtain

$$\begin{aligned} \frac{t}{w(q)} - \frac{t}{q} &= \frac{t}{q} \left[\frac{1}{w(\frac{1}{q})} - 1 \right] \\ &= \frac{t}{q} \cdot \sum_{n=1}^{\infty} P_n(\cos\gamma) \cdot \frac{1}{q^n} \\ &= \sin\theta \cdot \sum_{n=1}^{\infty} P_n(\cos\gamma) \cdot \frac{\sin^n\theta}{t^n} \\ &= \frac{\cos\gamma \cdot \sin^2\theta}{t} + \frac{1}{t^2} \sum_{n=2}^{\infty} \frac{\sin^{n+1}\theta}{t^{n-2}} \,. \end{aligned}$$

With

$$M(t) := \sum_{n=2}^{\infty} P_n(\cos\gamma) \frac{\sin^{n+1}\theta}{t^{n-2}} \,,$$

we have

$$\frac{t}{w(q)} - \frac{t}{q} = \frac{\cos\gamma \cdot \sin^2\theta}{t} + \frac{M(t)}{t^2} \,,$$

where M is a bounded function on $[1, +\infty)$, consequently the function $M(t)/t^2$ is integrable over this interval.

We introduce the following two notations:

$$S(y^s) := \int_0^{2\pi} d\phi \int_1^{\frac{h_0}{r_s}} \tilde{\mu}(r_s t, \phi) \frac{\cos\gamma \sin^2\theta}{t} \, dt = \int_0^{2\pi} d\phi \int_{r_s}^{h_0} \tilde{\mu}(\rho, \phi) \frac{\cos\gamma \sin^2\theta}{\rho} \, d\phi \tag{10}$$

and

$$\begin{aligned} R(y^s) := & \int_0^{2\pi} d\phi \int_0^1 \tilde{\mu}(r_s t, \phi) \left[\frac{t}{w(q)} - \frac{t}{q}\right] dt + \\ & + \int_0^{2\pi} d\phi \int_1^{\frac{h_0}{r_s}} \tilde{\mu}(r_s t, \phi) \cdot \left[\frac{t}{w(q)} - \frac{t}{q} - \frac{\cos\gamma \cdot \sin^2\theta}{t}\right] dt \, . \end{aligned} \tag{11}$$

Then we have

$$\frac{V_0(y^s) - V_0(x^0)}{r_s} = S(y^s) + R(y^s) \, .$$

We shall now show that

$$\lim_{r_s \to 0} R(y^s) = -\int_0^{2\pi} \tilde{\mu}(\phi) \cdot \sin^2\theta_0 \cdot \left\{1 + \cos\gamma_0 + \cos\gamma_0 \cdot \ln \frac{(1-\cos\gamma_0)\sin\theta_0}{2}\right\} d\phi \tag{12}$$

which shall complete the proof of Theorem 1.

Our assumption imply

$$\lim_{r_s \to 0} R(y^s) = \tilde{\mu}_0(\phi) \cdot \left\{\int_0^\infty \left[\frac{t}{w(q_0)} - \frac{t}{q_0}\right] dt - \cos\gamma_0 \cdot \sin^2\theta_0 \cdot \int_1^\infty \frac{st}{t}\right\} ,$$

where $q_0 = t/\sin\theta_0$. Let

$$I(\tau \sin\theta_0) := \int_0^{\tau \sin\theta_0} \left[\frac{t}{w(q_0)} - \frac{t}{q_0}\right] dt - \cos\gamma_0 \sin^2\theta_0 \int_1^{\tau \sin\theta_0} \frac{dt}{t} \quad .$$

By means of the coordinate transform $t \mapsto q_0$, where $q_0 = t/\sin\theta_0$, we obtain

$$\begin{aligned} I(\tau \sin\theta_0) &= \sin^2\theta_0 \cdot \left\{\int_0^\tau \left[\frac{q_0}{w(q_0)} - 1\right] dq_0 - \cos\gamma_0 \int_{1/\sin\theta_0}^\tau \frac{1}{q_0} \, dq_0\right\} \\ &= \frac{-2\tau\cos\gamma_0 + 1}{w(\tau) + \tau} + \cos\gamma_0 \cdot \ln \frac{w(\tau) + \tau - \cos\gamma_0}{\tau} - \\ & \quad - \{1 + \cos\gamma_0 \cdot \ln[(1-\cos\gamma_0)\sin\gamma_0]\} \quad . \end{aligned}$$

This implies

$$\lim_{\tau \to \infty} I(\tau \sin\theta_0) = -\sin^2\theta_0 \cdot \left\{1 + \cos\gamma_0 + \cos\gamma_0 \cdot \ln \frac{(1-\cos\gamma_0)\sin\theta_0}{2}\right\}$$

and hence equality (12) is proved. □

Proof of Theorem 2: As we noted before, the assumptions of this theorem imply $\forall \phi \in [0, 2\pi]$, $\tilde{\mu}_0(\phi) = \mu(x^0)$. Moreover we have $\theta_0 = \pi/2$, $\sin\theta_0 = 1$ and $\cos\gamma_0 = \sin\gamma_s \cos(\phi - \phi_s)$. Hence we obtain

$$\lim_{r_s \to 0} R(y^s) = -\mu(x^0) \cdot \int_0^{2\pi} \left\{1 + \cos\gamma_0 + \cos\gamma_0 \cdot \ln \frac{1 - cos\gamma_0}{2}\right\} d\phi \quad .$$

We note that for the function

$$X(\phi) := -\sin\theta_s \cdot \sin(\phi - \phi_s) \cdot \ln\frac{1-\cos\gamma_0}{2} - \cos^2\theta_s \cdot \int \frac{d\phi}{1-\cos\gamma_0}$$

we have

$$\begin{aligned} -\int_0^{2\pi} \left\{1 + \cos\gamma_0 + \cos\gamma_0 \cdot \ln\frac{1-\cos\gamma_0}{2}\right\} d\phi &= X(2\pi) - X(0) \\ &= -\cos^2\theta_s \cdot \int_0^{2\pi} \frac{d\phi}{1-\cos\gamma_0} \\ &= -2\pi \cdot \cos\theta_s \quad . \end{aligned}$$

[4]. It remains to show the existence of $\lim_{r_s\to 0} S(y^s)$. The additional assumptions of Theorem 2 imply $\tilde{\mu}(\rho,\phi) = \mu(x^0) + \rho \cdot \tilde{\mu}_1(\rho,\phi)$. Hence we may write

$$S(y^s) = \mu(x^0) \cdot \int_0^{2\pi} d\phi \int_{r_s}^{h_0} \frac{\cos\gamma \sin^2\gamma}{\rho} d\rho + \int_0^{2\pi} d\phi \int_{r_s}^{h_0} \tilde{\mu}_1(\rho,\phi) \cos\gamma \sin^2\theta \, d\rho \quad .$$

The second integral converges as $r_s \to 0$. By means of the equalities $\sin\theta = 1 + O(\rho)$ and $\cos\theta = \cos\gamma_0 + O(\rho)$ we obtain that the singular part of the first integral may be expressed by

$$\int_0^{2\pi} d\phi \int_{r_s}^{h_0} \frac{\cos\gamma_0}{\rho} d\rho = \sin\theta_s \int_0^{2\pi} \cos(\phi - \phi_s) \, d\phi \cdot \int_{r_s}^{h_0} \frac{d\rho}{\rho} = 0 \quad .$$

Hence $\lim_{r_s\to 0} R(y^s)$ exists. □

In the application of boundary element methods Γ is often a curvilinear triangle. We assume that the projection of Γ on the coordinate plan $y_3 = 0$ is a triangle denoted by Δ and that the point x^0 is the barycenter of Δ. Further we choose the positive number h_0 as the radius of the inscribed circle of Δ. Finally concerning the numerical aspects, let us restrict to several remarks. We start from

$$\frac{dV(x^0)}{ds} = \frac{dV_0(x^0)}{ds} + \frac{d}{ds} \iint_{\Gamma\setminus\Gamma_0} \frac{\mu(y)}{|y|} \, d\Gamma y := I_0 + I_1 \quad .$$

Since

$$I_1 = \frac{d}{ds} \iint_{\Gamma\setminus\Gamma_0} \frac{\mu(y)}{|y|} \, d\Gamma y = \iint_{\Gamma\setminus\Gamma_0} \mu(y) \frac{\cos(s,y)}{|y|^2} \, d\Gamma y$$

and

$$\cos(s,y) = \cos\gamma \quad ,$$

we obtain

$$I_1 = \int_0^{2\pi} d\phi \int_{h_0}^{\rho(\phi)} \tilde{\mu}(\rho,\phi) \frac{\cos\gamma}{r^2} \rho \, d\rho = \int_0^{2\pi} d\phi \int_{h_0}^{\rho(\phi)} \tilde{\mu}(\rho,\phi) \frac{\cos\gamma \sin^2\theta}{\rho} \, d\rho \quad ,$$

where $\rho = \rho(\phi)$ is the equation of the boundary of Δ. This shows that in case of Theorem 2 we have

$$\frac{dV(x^0)}{ds} = \lim_{r_s\to 0} \int_0^{2\pi} d\phi \int_{r_s}^{\rho(\phi)} \tilde{\mu}(\rho,\phi) \frac{\cos\gamma \sin^2\theta}{\rho} \, d\rho - 2\pi\mu(x^0) \cdot \cos\theta_s \tag{13}$$

This formula admits the numerical calculation of the derivative of V in x^0 in many situations. Usually the density function $\tilde{\mu}$ will be replaced by a polynomial $p_n(\rho, \phi)$ generated by Taylor's formula. If in addition $\Gamma = \Delta$ occures, then (13) will be reduced to the following simple formula

$$\frac{dV(x^0)}{ds} = \sin\theta_s \int_0^{2\pi} \cos(\phi - \phi_s)\, d\phi \int_0^{\rho(\phi)} \frac{p_n(\rho, \phi)}{\rho}\, d\rho - 2\pi\mu(x^0)\cos\theta_s \tag{14}$$

which admits an analytical evaluation of the integral.

REFERENCES

[1] M. H. ALIABADI, W. S. HALL and T. T. HIBBS: Exact singularity cancelling for the potential kernel in the boundary element method. Comm. Appl. Num. Meth. (1987), Vol.3, pp.123–128.

[2] M. H. ALIABADI and W. S. HALL, T. H. PHEMISTER: Taylor Expansions for Singular Kernels in the Boundary Element Method. Int. J. Num. Meth. Engrg. (1985), Vol.21, pp.2221–2236.

[3] E. ALLGOWER, K. GEORG, K. KALIK: Computation of weakly and nearly singular integrals over triangeles in $\mathbb{R}^3$. Preprint, August 1990; Colorado State University, Dept. Mathematics.

[4] I. S. GRADSHTEYN, I. M. RYZHIK: Table of integrals, series and products. Academic Press (1980)

[5] M. GUIGGIANI, A. GIGANTE: A general algorithm for multidimensional Cauchy principal value integrals in the boundary element method. J. of Appl. Mechanics 112 (1990) 906–915.

[6] K. HAYAMI and C. A. BREBBIA: A new coordinate transformation method for singular and nearly singular integrals over general curved boundary elements. In: Boundary Elements IX Vol. 1 (eds. C. A. Brebbia, W. L. Wendland, G. Kuhn), Springer-Verlag, Berlin, Heidelberg (1987) 375–399.

[7] R. KIESER: Über einseitige Sprungrelationen und hypersinguläre Operatoren in der Methode der Randelemente. Dissertation Fak. Mathematik, Universität Stuttgart, 1991.

[8] S. G. MIKHLIN and S. PRÖSSDORF: Singular integral operators. Springer-Verlag, Heidelberg (1986).

[9] C. SCHWAB and W. L. WENDLAND: 3-D BEM and numerical integration. In: Boundary Elements VII (eds. C. A. Brebbia, G. Maier) Vol. II, Springer-Verlag, Berlin (1985) 13.85–13.101.

[10] W. L. WENDLAND: Strongly elliptic boundary integral equations. In: The State of the Art in Numerical Analysis (A. Iserles and M. Powell eds.), Clarendon Press, Oxford (1987) 511–561.

The triangle-to-square transformation for finite-part integrals

Ralf Kieser
Universität Stuttgart, Mathematisches Institut A
Pfaffenwaldring 57, D-W-7000 Stuttgart 80

Summary

In this note we shall investigate the applicability of the so-called "triangle-to-square" coordinate transformation to the numerical evaluation of finite-part integrals arising in the context of 3D-BEM. Under certain symmetry hypotheses we show that the numerical computation of these integrals can be performed by using a tensor product of a one-dimensional finite-part integration formula and a one-dimensional formula for smooth integrands.
Numerical examples are presented.

Basic notions

Before presenting the method we shall briefly review the concept of a finite-part integral. (See, e.g., [4], §3.2.) Let A be a function on $(0, \varepsilon_0)$ which admits an expansion of the form

$$A(\varepsilon) = \sum A_j \varepsilon^{-\lambda_j} + A_0 \log \varepsilon + A + o(1), \quad \varepsilon \downarrow 0, \tag{1}$$

where the sum is finite and the λ_j are pairwise different with $\lambda_j \neq 0, \operatorname{Re} \lambda_j \geq 0$. Then the *finite part* of $A(\varepsilon)$ is defined by discarding the singular terms and then letting $\varepsilon \downarrow 0$:

$$\text{p.f.}\, A(\varepsilon) := A.$$

(Here, "p.f." stands for "partie finie", which is due to Hadamard [3].) It is easy to see that this definition is consistent in the sense that p.f. $A(\varepsilon)$ does not depend on the expansion. (See [4], Lemma 3.2.1.)

The finite part of a divergent integral does now appear as a special case : If, for example, $\phi \in C_0^\infty(\mathbb{R})$ (i.e. smooth with compact support) and $k \in \mathbb{N}_0$, we have

$$\begin{aligned}\int_\varepsilon^\infty x^{-1-k}\phi(x)dx &= -\frac{1}{k!}\int_0^\infty (\log x)\phi^{(k+1)}(x)dx + \sum_{j=1}^{k} A_j\varepsilon^{-j} + \\ &\quad + A_0 \log\varepsilon + \frac{\phi^{(k)}(0)}{k!}\sum_1^k \frac{1}{j!} + o(1), \quad \varepsilon \downarrow 0, \end{aligned} \tag{2}$$

where the A_j only depend on k and on derivatives of ϕ at 0. (Relation (2) follows from integrating by parts $k+1$ times and then applying Taylor's formula in order to expand the terms $\phi^{(j)}(\varepsilon)$ around the origin.) Then we set

$$\oint_0^\infty x^{-1-k}\,\phi(x)\,dx := \text{p.f.} \int_\varepsilon^\infty x^{-1-k}\phi(x)\,dx \tag{3}$$

and refer to (3) as a *finite-part integral.*

The n-dimensional case is handeled by introducing polar coordinates: Assume that the function u is locally integrable on $\mathbb{R}^n\backslash 0$ and (positively) homogeous of degree $-n-k$, $k \in \mathbb{N}_0$, i.e.

$$u(\lambda x) = \lambda^{-n-k} u(x); \quad x \in \mathbb{R}^n\backslash 0,\ \lambda > 0.$$

Then we define for $\phi \in C_0^\infty(\mathbb{R}^n)$

$$⨍ u\phi\, dx := \int_{|\omega|=1} u(\omega)\ ⨍_0^\infty r^{-1-k}\phi(r\omega)\, dr\, d\omega = \text{p.f.} \int_{|x|\geq\varepsilon} u\phi\, dx, \tag{4}$$

where $d\omega$ denotes the surface element on the unit sphere. In the context of distribution theory, (4) can be regarded as an extension of u (considered as a distribution on $\mathbb{R}^n\backslash 0$) to a distribution on $\mathbb{R}^n$. Note that the Cauchy principal value appears as a special case of (4) when $k = 0$ and the integral of u on the unit sphere (and thus the divergent part) vanishes.

The case of finite-part integrals on finite integration domains can be dealt with analogously or, alternatively, reduced to the previous case by introducing a cut-off function. (Note that a certain smoothness of ϕ is required in a neighbourhood of the origin.) For example, if $\lambda > 0$, we have

$$⨍_0^\lambda x^{-1} dx = \text{p.f.} \int_\varepsilon^\lambda x^{-1}\, dx = \text{p.f.}\,\{\log\lambda - \log\varepsilon\} = \log\lambda.$$

This also exemplifies the well-known fact that the ordinary change of variables formula (here: scaling by λ) is in general *not* valid for finite-part integrals. For a linear bijection $L : \mathbb{R}^n \to \mathbb{R}^n$ the transformation rule for (4) is

$$\begin{aligned} ⨍ u(Ly)\,\phi(Ly)\,|\det L|\, dy &= ⨍ u(x)\phi(x)\, dx - \\ &- \sum_{|\alpha|=k} \partial^\alpha \phi(0) \int_{|\omega|=1} \omega^\alpha u(\omega) \log R(\omega)\, d\omega/\alpha!, \end{aligned} \tag{5}$$

where $\phi \in C_0^\infty(\mathbb{R}^n)$ and R denotes the polar coordinate representation of the surface $\{Ly : |y| = 1\}$. (See [6], §2.2, for a proof.)

Finite-part integrals in BEM

In the context of BEM, solutions to boundary value problems (and other relevant physical quantities) are represented in terms of *Green potentials* of the form

$$U(X) = \int_\Gamma K(X,Y)\, f(Y)\, do(Y), \quad X \subset \Omega, \tag{6}$$

where Γ is the boundary of a domain $\Omega \subset \mathbb{R}^n$ and K can be regarded as the (distributional) kernel of a pseudo-differential operator in $\mathbb{R}^n$. A well-known example is the normal derivative of the double-layer potential to the Laplace operator in $\mathbb{R}^3$,

$$4\pi K(X,Y) := \partial_{\nu(X)}\partial_{\nu(Y)} \frac{1}{|X-Y|} = \frac{\langle \nu(X), \nu(Y)\rangle}{|X-Y|^3} + \cdots, \tag{7}$$

where $\langle \cdot, \cdot \rangle$ is the usual scalar product, and ν denotes (an extension of) the exterior unit normal field on Γ. Taking the trace $X \to \Gamma$ in (6) defines (for Γ being sufficiently smooth) an operator in Γ which can be represented in local coordinates as the sum of a differential operator acting on u and a finite-part integral. (Depending on the singularity of K for $X = Y$, this integral may or may not exist in the ordinary sense.) For a rather general class of kernels K it can be shown that the correponding finite-part integrals do not depend on the local coordinates chosen on Γ if Γ is smooth near the singular point X. (Cf. [7] and [2], Lemma 8.1.11, for special cases, and [6], §4.3, for the general case.) Note that this is in contrast to the behaviour of *general* finite-part integrals with respect to a change of variables (cf. (5)).

Decomposition into homogeneous terms

In (6) we introduce a smooth and regular local representation $\chi : D \to \Gamma$ of Γ, $D \subset \mathbb{R}^2$ being an open set. This yields a finite-part integral of the form

$$I := \oint_D K(\chi(x), \chi(y))\, f(\chi(y))\, J_\chi(y)\, dy,$$

where J_χ denotes the Jacobian. Assuming $0 \in D$, $X = \chi(0) = 0$ without loss of generality, we get with $v := K(0, \cdot)$, $\phi := (f \circ \chi) J_\chi$

$$I = \oint_D v(\chi(y)) \cdot \phi(y)\, dy, \tag{8}$$

where ϕ is smooth on $\bar{D}$. Regarding v we make the following

Assumption: For any $N \in \mathbb{N}$ we can write

$$v \circ \chi = u_0 + u_1 + \cdots + u_{N-1} + R_N \quad \text{on} \quad D, \tag{9}$$

where for some $k \in \mathbb{N}_0$ the following conditions are satisfied:

- u_j is homogeneous on $\mathbb{R}^2 \backslash 0$ of degree $-2 - k + j$ for $j \in \mathbb{N}_0$;
- u_j has parity $k - j + 1$, i.e. $u_j(-y) = (-1)^{k-j+1}\, u_j(y)$ for $y \in \mathbb{R}^2 \backslash 0$;
- $D^\alpha R_N(y) = O(|y|^{-2-k+N-|\alpha|})$ when $y \to 0$ for any multi-index $\alpha \in \mathbb{N}_0^2$.

This assumption is satisfied if v is homogeneous of degree $-2-k$ in $\mathbb{R}^3 \backslash 0$ and has parity $k+1$. In this case, the u_j can be computed using Taylor's formula by first expanding χ around the origin and then expanding v around $\chi'(0)y$:

$$v(\chi(y)) = v(0 + \chi'(0)y + \cdots) = v(\chi'(0)y) + \cdots$$

If, for example, $v(Y) := |Y|^{-3}$ we get

$$\frac{1}{|\chi(y)|^3} = \frac{1}{|\chi'(0)y|^3} - 3 \cdot \sum_{|\alpha|=2} \frac{< \chi'(0)y, \partial^\alpha \chi(0) >}{|\chi'(0)y|^5} \cdot \frac{y^\alpha}{\alpha!} + O(\frac{1}{|y|}).$$

Note that the Assumption remains valid if $v \circ \chi$ is multiplied by a function which is smooth on D and can thus be expanded in polynomial terms by Taylor's formula. Hence, the Assumption is valid for the right hand side of (7), and the same is true for the corresponding terms in the case of linearized elasticity (which arise as matrix entries when applying the traction operator to the kernel matrix of the double-layer potential).

As is well-known, the device of decomposition into homogenous terms can be used to compute (8) in a semi-analytic way: After introducing polar coordinate on D the (finite-part) integration in radial direction becomes trivial for the homogeneous terms, whilst they are piecewise smooth with respect to the angular variable if the boundary of D is piecewise smooth. The remaining integrations (including the term with R_N) can then be performed numerically in a standard way. If, for example, $\theta \to R(\theta)$ is the polar coordinate representation of the boundary of D, we have

$$\begin{aligned} \not\!\!\int_D |\chi'(0)y|^{-3} &= \int_0^{2\pi} \not\!\!\int_0^{R(\theta)} r^{-2} \left|\chi'(0)\begin{pmatrix}\cos\theta\\ \sin\theta\end{pmatrix}\right| dr\, d\theta = \\ &= -\int_0^{2\pi} R(\theta)^{-1} \cdot \left|\chi'(0)\begin{pmatrix}\cos\theta\\ \sin\theta\end{pmatrix}\right| d\theta. \end{aligned}$$

In contrast to this, our aim here is to design a quadrature method which (under additional assumptions on D) allows to compute (8) "directly" (i.e. without analytic calculations to be done in advance).

Triangle-to-square transformation

Let T and Q denote the standard triangle $T := \{(x_1, x_2) \in \mathbb{R}^2 : 0 < x_2 < x_1 < 1\}$ and the unit square $S := \{(\xi, \eta) \in \mathbb{R}^2 : 0 < \xi < 1, 0 < \eta < 1\}$, respectively. The transformation

$$x_1 = \xi\,, \quad x_2 = \xi\eta\,, \quad dx = \xi\, d(\xi, \eta)$$

is one-to-one between T and S and brings out a factor ξ in the differential, which cancels the singularity in the case of weakly singular integrals (see [1], [5], [8]). In the finite-part case there appears an additional term in the transformation:

Lemma: *If $u : \mathbb{R}^2\backslash 0 \to \mathbb{R}$ is homogeneous of degree $-2-k$ for some $k \in \mathbb{N}_0$ and ϕ is smooth on $\bar{T}$ then*

$$\begin{aligned} \not\!\!\int_T u\phi\, dx = \int_0^1 u(1,\xi)\ \not\!\!\int_0^1 \xi^{-1-k}\phi(\xi,\xi\eta)\, d\xi\, d\eta + \\ + \int_0^1 u(1,\eta) \log\sqrt{1+\eta^2}\frac{1}{k!}(\partial_1 + \eta\partial_2)^k \phi(0,0)\, d\eta\,. \end{aligned} \tag{10}$$

For a proof see [6], §5.2.

In this form the transformation (10) is of little practical interest since the evaluation of the additional term on the right hand side would require analytic work (which we want to avoid). However, it is easily seen that this term cancels out if (10) is performed simultaneously for T and $-T := \{x : -x \in T\}$ and if v has parity $k+1$. Combining this observation with (9) and (5) yields the following

Theorem: *Let $T' \subset \mathbb{R}^2$ be a triangle having the origin as a corner, and let $L : T \to T'$ be linear and one-to-one. Assume that $D \supset B := -\bar{T}' \cup \bar{T}'$ is open and $\chi : D \to \mathbb{R}^3$ is smooth and regular with $\chi(0) = 0$. If v is a smooth function on $\mathbb{R}^3\backslash 0$ such that the above Assumption is satisfied, and if ϕ is smooth on D, then we have with $g := (v\circ\chi\circ L)\cdot(\phi\circ L)$*

$$\begin{aligned} &⨍_B (v \circ \chi)\phi \, dy = \\ &\qquad |\det L| \int_0^1 ⨍_{-1}^1 \xi^{-1-k} \left\{\xi^{1+k}|\xi| g(\xi,\xi\eta)\right\} d\xi \, d\eta, \end{aligned} \tag{11}$$

and the function in $\{\cdots\}$ is smooth on $[-1,1] \times [0,1]$.

For a proof see [6], §5.3.

The right hand side of (11) can now be computed numerically by combining a one-dimensional finite-part integration formula on $[-1,1]$ for the "weight" ξ^{-1-k} and a one-dimensional formula on $[0,1]$ for smooth integrands. By a subdivision, formula (11) can be applied when the integration domain is a rectangle with center at the origin. If this is not the case we can subdivide the domain of integation into a rectangle centered at the origin and a domain where the integrand is smooth.

Numerical examples

In what follows, formula (11) will be applied to some finite-part integrals of the form

$$⨍_{S_h} (v \circ \chi)\phi \, dy \quad \text{with} \quad S_h := \{y \in \mathbb{R}^2 : |y_1|, |y_2| < h\} .$$

Example 1: $v(Y) := |Y|^{-3}$, $\chi(y) := (y, 0)$, $k = 1$, $\phi := 1$.

Example 2: $v(Y) := 4\pi K(X,Y)$, K as in (7) , with the half sphere
$\Gamma := \{Y \in \mathbb{R}^3 : |Y| = 1, Y_1 > 0\}$ given in polar coordinates by
$Y = \chi(y) := (\cos y_1 \cos y_2, \sin y_1 \cos y_2, \sin y_2)$, $Y \in S_{\pi/2}$;
$X := \chi(0) = (1,0,0)$, $k = 1$, $\phi(y) := \cos y_2$. (Note that $\phi\, dy$ is the surface element.)

Example 3: $v(Y) := |Y|^{-3}$, $\chi(y) := (y_1, y_2, y_1 y_2)$, $k = 1$, $\phi := 1$.

The results are given in the following table.

example no.	# nodes "$\xi \otimes \eta$"	exact value	numerical result sngl. prec.	dbl. prec.
1	2×3	-5.6568527	-5.75 ...	
	2×6		-5.6562 ...	
	2×9		-5.65685 ...	
2	6×5	-2.2214413	-2.20 ...	
	8×6		-2.2210 ...	
	10×7		-2.22150 ...	-2.22148 ...
	12×12		-2.22138 ...	-2.22142 ...
3	4×4	-12.000000	-11.96 ...	
	8×6		-12.0002 ...	
	8×9		-12.00002 ...	
	12×12		-11.9998 ...	-12.000000

The integration with respect to ξ is performed using Newton-Cotes-type formulae for the "weight" ξ^{-2} on $[-1, 1]$, whereas the integration with respect to η is done using Gaussian rules. Values in double precision are given only if there is a significant improvement compared to single precision. The numerical results were obtained using FORTRAN on the COMPAREX 8/89 at the Rechenzentrum der Universität Stuttgart.

Conclusion

If the proposed method is applicable (i.e. essentially: if the singular point is an *interior* point of the coordinate patch), then satisfactory accuracy can be achieved without performing analytic integrations but at the expense that the required number of nodes is not small.

References

[1] M. G. Duffy. Quadrature over a pyramid or cube of integrands with a singularity at a vertex. *SIAM J. Numer. Anal.*, 19:1260–1262, 1982.

[2] W. Hackbusch. *Integralgleichungen: Theorie und Numerik.* Teubner, Stuttgart, 1989.

[3] J. Hadamard. *Lectures on Cauchy's Problem.* Yale University Press, 1923.

[4] L. Hörmander. *The Analysis of Linear Partial Differential Operators*, volume 1. Springer, Berlin, 1983.

[5] C. G. L. Johnson and L. R. Scott. An analysis of quadrature errors in second-kind boundary integral methods. *SIAM J. Numer. Anal.*, 26:1356–1382, 1989.

[6] R. Kieser. *Über einseitige Sprungrelationen und hypersinguläre Operatoren in der Methode der Randelemente.* PhD thesis, Universität Stuttgart, 1991.

[7] P. A. Martin and F. J. Rizzo. On boundary integral equations for crack problems. *Proc. R. Soc. Lond. A*, 421:341–355, 1989.

[8] C. Schwab and W. L. Wendland. On numerical cubatures of singular surface integrals in boundary element methods. (In Vorbereitung.)

NUMERICAL SOLUTION OF THE OBLIQUE DERIVATIVE PROBLEM IN $\mathbb{R}^3$ USING THE GALERKIN–BUBNOV–METHOD: NUMERICAL INTEGRATION, SOLUTION OF THE LINEAR SYSTEM OF EQUATIONS AND THE USE OF VECTOR PIPELINE MACHINES

R. Klees

Geodetic Institute, University of Karlsruhe (TH)

P.O. Box 6980, D–7500 Karlsruhe 1, Germany

SUMMARY

We consider the classical oblique boundary–value problem in $\mathbb{R}^3$ which arises after linearisation of the fixed gravimetric boundary–value problem of Physical Geodesy. The indirect formulation using the potential of the single layer leads to a strongly singular integral equation for the single layer density on the earth surface. We solve this integral equation by using the Galerkin–Bubnov–discretization with piecewise constant test and trial functions. Each main diagonal element is defined as the sum of a single regular integral and a cauchy–singular double integral over the boundary element whereas the off–diagonal elements are regular double integrals. We develop efficient cubature formulas for the determination of all the integrals on vector pipeline machines. Moreover we use a direct method for the solution of the linear system of equations adapted in an optimal sense to the vector pipeline machine and based on the LU–factorization using in principal the left–looking method of gaussian elimination. Besides this we examine different methods for the pointwise computation of the solution of the boundary–value problem showing that efficient computation is possible even if the computation point is very near by the earth's surface.

INTRODUCTION

The determination of the earth's gravitational potential belongs to the main tasks in Physical Geodesy. There are ranges of application for very high resolution gravity fields not only in Physical Geodesy but also in Geophysics, Oceanography and Geodynamics, e.g. the definition of a unique hight system, the improved determination of satellite orbits, the determination of the sea surface topography and of deformations of the earth's crust and a deeper understanding of the processes both in the earth's lithosphere and mantle.

Up to now mainly free boundary value problems have been considered in Physical Geodesy. But in view of the great progress in satellite geodesy, especially in satellite altimetry and satellite positioning techniques and the continual development of high resolution digital terrain models we can assume that the earth's surface is a priori known by strictly geometric means so that we only have to determine the gravitational potential in the earth's outer space. Using the magnitude of the gravity gradient ("gravity values") as boundary data, which can be measured on the real earth's surface, we get the so called fixed gravimetric boundary–value problem of Physical Geodesy, a nonlinear boundary–value problem for the Laplace operator. For a detailed mathematical analysis of this problem see [1] and the references given there.

FORMULATION OF THE PROBLEM

Using the implicit function theorem of Hildebrand and Graves [2] or a perturbation technique [1] we can solve this nonlinear problem iteratively having to solve a classical oblique boundary–value problem in each step. We consider one step in this iteration process and formulate this boundary–value problem as follows: Given the earth's surface Γ and a continuous covering of Γ with gravity values g; let us denote by V the earth's gravitational potential, V_0 an approximate value of V, Ω the centrifugal potential, $\boldsymbol{l}$ the unit vector in direction of $\nabla(V_0+\Omega)$, Ω_a the earth's outer space and v the improvement of the approximation V_0 in the current iteration step so that $V_0 + v$ is the approximation after the step under consideration. Then

$$\begin{aligned} &\Delta v(\boldsymbol{x}) = 0 \quad \text{for } \boldsymbol{x} \in \Omega_a \\ &< \boldsymbol{l}(\boldsymbol{x}), \nabla v(\boldsymbol{x}) > = g(\boldsymbol{x}) - |\nabla(V_0+\Omega)(\boldsymbol{x})| \quad \text{for } \boldsymbol{x} \in \Gamma \\ &v(\boldsymbol{x}) = O(|\boldsymbol{x}|^{-1}) \quad \text{for } |\boldsymbol{x}| \to \infty \ , \end{aligned} \tag{1}$$

TRANSFORMATION INTO A BOUNDARY INTEGRAL EQUATION

For the numerical solution of problem (1) we transform it into a boundary integral equation by using the potential of a single layer as representation formula for the unknown potential v

$$v(\boldsymbol{x}) := \frac{1}{4\cdot\pi}\cdot\int_{\boldsymbol{y}\in\Gamma} \frac{f(\boldsymbol{y})}{|\boldsymbol{x}-\boldsymbol{y}|}\, d\Gamma(\boldsymbol{y})\,. \tag{2}$$

If we put equation (2) into the boundary condition (1) and consider the well known jump relations for the single layer potential if a point $\boldsymbol{x}_0 \in \Omega_a$ approaches the point $\boldsymbol{x} \in \Gamma$ along an axis $\boldsymbol{l}$ through $\boldsymbol{x}$ which is not tangential to the boundary Γ we get a strongly singular integral equation of the second kind for the single layer density f on Γ:

$$-\frac{1}{2}\cdot f(\boldsymbol{x})\cdot\cos[\boldsymbol{n}(\boldsymbol{x}),\boldsymbol{l}(\boldsymbol{x})] + \frac{1}{4\cdot\pi}\,\text{p.v.}\int_{\boldsymbol{y}\in\Gamma} \frac{< \boldsymbol{l}(\boldsymbol{x}), \boldsymbol{y}-\boldsymbol{x} >}{|\boldsymbol{x}-\boldsymbol{y}|^3}\cdot f(\boldsymbol{y})\, d\Gamma(\boldsymbol{y}) = h(\boldsymbol{x}), \tag{3}$$

where $\boldsymbol{x} \in \Gamma$, $h(\boldsymbol{x}) := g(\boldsymbol{x}) - |\nabla(V_0+\Omega)(\boldsymbol{x})|$. $\boldsymbol{n}(\boldsymbol{x})$ denotes the unit normal vector pointing into Ω_a and $\boldsymbol{l}(\boldsymbol{x})$ denotes the unit vector in the direction of the gradient $\nabla(V_0+\Omega)$ each taken at a point $\boldsymbol{x} \in \Gamma$. For (3) we write shortly

$$(\mathbf{A}f)(\boldsymbol{x}) = h(\boldsymbol{x}) \quad , \boldsymbol{x} \in \Gamma. \tag{4}$$

The properties of the operator **A** are based on the explicit representation with regard to local parametrizations of the boundary surface Γ and its interpretation both as a pseudodifferential operator of order zero and a singular integral operator.

DISCRETIZATION OF THE BOUNDARY INTEGRAL EQUATION

In order to transform the boundary integral equation (3) into a finite dimensional linear system of equations we discretize the boundary into finite elements Γ_i and approximate the unknown single layer density using piecewise constant finite elements with regard to parameter representations

$$f(\boldsymbol{x}) = \sum_{i=1}^{n} \alpha_i \cdot \mu_i(\boldsymbol{x}) \ , \tag{5}$$

where n denotes the number of finite elements and

$$\mu_i(\boldsymbol{x}) = \begin{Bmatrix} 1 & \text{for } \boldsymbol{x} \in \Gamma_i \\ 0 & \text{else} \end{Bmatrix} \ . \tag{6}$$

Using the Galerkin–Bubnov–method to discretize the boundary integral equation leads to a linear system of equations $A \cdot \boldsymbol{\alpha} = \boldsymbol{h}$ with $A := (a_{ik})$, $\boldsymbol{\alpha} := (\alpha_k)$ and $\boldsymbol{h} := (h_i)$, where the coefficients are defined by

$$a_{ik} := -2\pi \cdot \delta_{ik} \cdot \int_{\boldsymbol{x}\in\Gamma_i} \cos[\boldsymbol{n}(\boldsymbol{x}), \boldsymbol{l}(\boldsymbol{x})] \, d\Gamma(\boldsymbol{x}) + \int_{\boldsymbol{x}\in\Gamma_i} \int_{\boldsymbol{y}\in\Gamma_k} \frac{\cos[\boldsymbol{l}(\boldsymbol{x}), \boldsymbol{y}-\boldsymbol{x}]}{|\boldsymbol{y}-\boldsymbol{x}|^2} \, d\Gamma(\boldsymbol{y}) d\Gamma(\boldsymbol{x}) \tag{7}$$

$$h_i = 4\pi \cdot \int_{\boldsymbol{x}\in\Gamma_i} (g - |\nabla(V_0+\Omega)|)(\boldsymbol{x}) \, d\Gamma(\boldsymbol{x}) \quad \text{for } i,k = 1,\ldots,n \ ,$$

and δ_{ik} denotes the Kronecker symbol.

NUMERICAL INTEGRATION

Computing the elements of the linear system of equations is the major difficulty in boundary element methods. In our case we have to compute both single and double integrals over the boundary elements Γ_i being parts of the earth's surface, where integrals of regular, weakly singular and strongly singular type occur as well as integrals which are regular in the mathematical sense but behave like singular integrals ("quasi–singular integrals"). In the following we have to restrict ourselves to some aspects of the whole difficulty. For more details, especially a detailed comparison of several numerical integration methods for single and double weakly singular and strongly singular integrals over two–dimensional surfaces in $\mathbb{R}^3$ and their pros and cons see [1].

We consider first the main diagonal elements of the linear system of equations. They are defined as the sum of a regular single integral

$$a_{ii,r} = -2\pi \cdot \int_{\boldsymbol{x}\in\Gamma_i} \cos[\boldsymbol{n}(\boldsymbol{x}), \boldsymbol{l}(\boldsymbol{x})] \, d\Gamma(\boldsymbol{x}) \quad , i=1,\ldots,n \tag{8}$$

and a double integral whose inner integral is to be understood in the sense of Cauchy's principal value (see 7). For the computation of the regular single integrals defined by equation (8) we use simple composed product formulas of Gauss–Legendre type. For the computation of the unit normal vectors $\boldsymbol{n}(\boldsymbol{x})$ the use of Overhauser–splines [3] based on rectangular elements has to be proved a very powerful method, especially on pipelined computers [see 1 for more details]. For an appropriate choice of the number of knots the cubatures are even exact, the only error source coming from the approximation of Γ.

In order to evaluate the double integrals

$$\int_{\boldsymbol{x}\in\Gamma_i} \int_{\boldsymbol{y}\in\Gamma_k} \frac{\cos[\boldsymbol{l}(\boldsymbol{x}), \boldsymbol{y}-\boldsymbol{x}]}{|\boldsymbol{y}-\boldsymbol{x}|^2} \, d\Gamma(\boldsymbol{y}) d\Gamma(\boldsymbol{x}) \quad , i,k = 1,\ldots,n \tag{9}$$

we represent the inner integral as the sum of a strongly singular integral having a kernel with a simpler structure than the original kernel, and a weakly singular remaining term and compute them separately. For

$$\Gamma_i = \{ \mathbf{x} \in \mathbb{R}^3 : \mathbf{x} = \varphi_i(x,y) = (x,y,z(x,y))^T,\ (x,y) \in S_i \subset \mathbb{R}^2 \} \quad , \tag{10}$$

where x,y,z denote rectangular cartesian coordinates in $\mathbb{R}^3$, we get

$$\frac{\cos[\mathbf{l}(\mathbf{x}),\mathbf{y}-\mathbf{x}]}{|\mathbf{y}-\mathbf{x}|^2}\, d\Gamma(\mathbf{y}) = \frac{\chi(\mathbf{x},\psi)}{\rho^2}\, d\eta + \frac{\vartheta(\mathbf{x},\rho,\psi)}{\rho}\, d\eta \quad , \tag{11}$$

where ξ and η are the images of $\mathbf{x}$ and $\mathbf{y}$, resp., under the transformation φ_i^{-1} and ρ,ψ denote the polar coordinates of the point η with respect to ξ. It can be shown that [1]

$$\lim_{\epsilon\to 0^+} \int_{\mathbf{y}\in\Gamma_i\setminus|\mathbf{x}-\mathbf{y}|<\epsilon} \frac{\cos[\mathbf{l}(\mathbf{x}),\mathbf{y}-\mathbf{x}]}{|\mathbf{x}-\mathbf{y}|^2}\, d\Gamma(\mathbf{y}) = \lim_{\tau\to 0^+} \int_{\eta\in S_i\setminus\rho<\tau} \rho^{-2}\cdot\chi(\mathbf{x},\psi)\, d\eta + \int_{\eta\in S_i} \frac{\vartheta(\mathbf{x},\rho,\psi)}{\rho}\, d\eta \tag{12}$$

and $\vartheta(\mathbf{x},\rho,\psi) = O(1)$ for $\rho \to 0$ the last integral in (12) being weakly singular. Moreover, the function $\vartheta(\mathbf{x},\rho,\psi)$ is discontinuous at the point $\rho=0$ and shows strong oszillations within the immediate adjacency of $\mathbf{x} = \varphi_i(\xi)$. The characteristic χ fulfils the Mikhlin–condition and we get for the first integral in (12)

$$\lim_{\tau\to 0,\tau>0} \int_{\eta\in S_i\setminus\rho<\tau} \rho^{-2}\cdot\chi(\mathbf{x},\psi)\, d\eta = \int_0^{2\pi} \chi(x,\psi)\cdot\ln[R(\psi)]\, d\psi \quad , \tag{13}$$

where $R(\psi)$ denotes the equation of the boundary ∂S_i of S_i in polar coordinates with respect to the pole ξ. The integral (13) can be computed analytically and it has been proved in [1] that the resulting function $H(\mathbf{x})$ has a logarithmic singularity on the whole boundary $\partial\Gamma_i = \varphi_i(\partial S_i)$ of the boundary element $\Gamma_i = \varphi_i(S_i)$. Therefore, we compute each of the integrals (9) as the sum of a strongly singular part $a_{ii,c}$ and a weakly singular part $a_{ii,s}$ defined by:

$$a_{ii,c} := \int_{\mathbf{x}\in\Gamma_i} H(\mathbf{x})\, d\Gamma(\mathbf{x}) \quad , i=1,\dots,n \tag{14}$$

$$a_{ii,s} := \int_{\mathbf{x}\in\Gamma_i} \int_{\eta\in S_i} \frac{\vartheta(\mathbf{x},\rho,\psi)}{\rho}\, d\eta\, d\Gamma(\mathbf{x}) \quad , i=1,\dots,n \, . \tag{15}$$

Due to the logarithmic singularity of the kernel $H(\mathbf{x})$ on the whole boundary $\partial\Gamma_i$ resp. ∂S_i we have to use special cubature formulas for the computation of the integrals (14). Due to their greater flexibility we prefer instead of gaussian cubature formulas with a logarithmic weight function special one–dimensional parameter transformations of the kind

$$1 \pm x = y^k,\ k > 1 \quad . \tag{16}$$

Using (16), integrals of the form

$$\int_{-1}^{-1} \ln(1\pm x)\, dx \tag{17}$$

transform into

$$k \cdot \int_0^{2^{1/k}} y^{k-1} \cdot \ln(y^k)\, dy \quad , \tag{18}$$

where the new kernel is well behaved for $y \to 0$ so that we can use Gauss–Legendre quadrature formulas for the efficient computation of integrals of type (17). In order to demonstrate the efficiency of the parameter transformation (16) we compute numerically integrals of type (17) using different values of k and different numbers of knots for the Gauss–Legendre quadratures. Table 1 shows that the best results are obtained for exponents

$$4 \leq k \leq n \quad , \tag{19}$$

where n denotes the number of knots of the quadrature formula. Other parameter transformations are possible and have been discussed in the literature, e.g. [4, 5, 6]. But several test computations show that all these methods have less accuracy for comparable numbers of knots, are more time–consuming, more difficult to implement and more sensitive against round–off errors, especially IMT– and ERF–rule. Moreover the source code is in some cases more difficult to adapt to pipelined computers to get high performance. Table 2 shows some results which are typical for the numerical computation of the strongly singular parts $a_{ii,c}$.

Table 1 Numerical computation of (17) for different exponents k.
Here: Relative cubature errors.

k	number of knots				
	2	5	8	10	20
1	1.0 (± 0)	2.1 (−2)	8.8 (−3)	5.7 (−3)	1.5 (−3)
2	3.1 (−2)	1.1 (−3)	2.0 (−4)	8.4 (−5)	5.7 (−6)
3	2.3 (−2)	1.2 (−4)	8.3 (−6)	2.3 (−6)	4.0 (−8)
4	4.9 (−2)	2.2 (−5)	5.8 (−7)	1.0 (−7)	4.7 (−10)
5	1.9 (−1)	6.6 (−6)	6.3 (−8)	7.1 (−9)	8.1 (−12)
8	4.4 (−1)	2.8 (−6)	6.8 (−10)	1.8 (−11)	2.7 (−16)
10	4.0 (−1)	3.0 (−5)	1.6 (−10)	1.3 (−12)	1.2 (−19)

Table 2 Numerical computation of the strongly singular part $a_{11,c}$ of the main diagonal coefficient a_{11}. $5n^2$: Total number of knots.
Here: Relative cubature errors.

$5n^2$	Exponent k				
	1	2	4	6	8
180	−5.2 (−2)	3.1 (−3)	2.0 (−3)	1.6 (−3)	3.9 (−4)
500	2.0 (−2)	3.6 (−4)	7.0 (−5)	3.4 (−5)	3.0 (−4)
2000	5.3 (−3)	1.9 (−5)	2.1 (−6)	2.1 (−6)	——

For the computation of the weakly singular parts $a_{ii,s}$ (15) several methods are available; their use depends on the properties of the kernel and the requirements with regard to accuracy and CPU–time. Some methods are based on the approximation of the integrand by functions which can be integrated analytically [7], others use gaussian type formulas the singularity being taken into the weight function, and the knots and weights are determined so that the resulting cubature formula is exact for polynomials up to a certain degree [8, 9]. Other methods use parameter transformation techniques [10, 11], taylor expansions to isolate the singularity [12] or other special cubature formulas [9, 13]. Polar coordinates or triangle coordinates or other special parameter transformations reduce the order of the singularity by one so that weakly singular integrals are transformed into regular ones and standard cubature formulas for the transformed integral can be used [3, 14, 15, 16]. But in most cases the singularity is ignored and simple Gauss–Legendre product formulas are used.

We investigate several of the methods mentioned above ($\{\cdot\}$ denotes the method in Table 3): Simple Gauss–Legendre product formulas {1}, the cubic transformation of Telles [4] applied to the cartesian coordinates {2}, polar coordinates {3}, two–dimensional IMT–transformation [17] {4}, triangle coordinates {5}, the angular transformation of Hayami & Brebbia [18] {6} and a combination of polar coordinates, the cubic transformation of Telles applied to the polar distance and the angular transformation of Hayami & Brebbia applied to the polar angle {7}. Here we have to confine ourselves to the numerical calculation of the <u>inner</u> integral of (15) that being sufficient for getting an impression of the efficiency of some of the methods mentioned before. Table 3 shows the results. Ignoring the singularity by using simple Gauss–Legendre product formulas is a

Table 3 Numerically evaluated values of the inner integral of (15) and their relative accuracy
Nominal value (12 significant figures): 0.220687092097

	number of knots					
	16	36	64	100	256	400
1	0.194306 1.2 (–1)	0.186200 1.6 (–1)	0.290992 3.2 (–1)	0.209102 5.2 (–2)	0.213228 3.4 (–2)	0.248721 1.3 (–1)
2	0.235850 6.9 (–2)	0.222995 1.0 (–2)	0.220894 9.4 (–4)	0.220587 4.5 (–4)	0.222123 6.5 (–3)	0.220830 6.5 (–4)
3	0.218510 9.9 (–3)	0.220792 4.8 (–4)	0.220822 6.1 (–4)	0.220738 2.3 (–4)	0.2206882 5.1 (–6)	0.22068716 3.2 (–7)
4	0.265179 2.0 (–1)	0.217951 1.2 (–2)	0.215384 2.4 (–2)	0.218997 7.7 (–3)	0.220802 5.2 (–4)	0.220744 2.6 (–4)
5	0.228624 3.6 (–2)	0.216062 2.1 (–2)	0.221943 5.7 (–3)	0.220653 1.5 (–4)	0.220675 5.5 (–5)	0.2206896 1.1 (–5)
6	0.219674 4.6 (–3)	0.220866 8.1 (–4)	0.220673 6.4 (–5)	0.2206857 6.3 (–6)	0.22068711 8.2 (–8)	.2206870922 8.0 (–10)
7	0.224529 1.7 (–2)	0.220334 1.6 (–3)	0.220674 5.9 (–5)	0.2206881 4.6 (–6)	0.22068712 1.3 (–7)	.2206870923 8.0 (–10)

very fast method and, moreover, the source code is perfectly well vectorizable, but the accuracy is completely insufficient; the same holds for weighted gaussian formulas (not contained in Table 3, see [1]). The two–dimensional IMT–transformation produces smaller errors than the simple Gauss–Legendre product formula but needs a lot of knots to get a certain accuracy and is very time–consuming. The cubic parameter transformation proposed by Telles yields reasonable accuracy for acceptable numbers of knots. In relation to all other methods it does not require the partition of the domain of integration into four triangles with the singularity as one common corner point. But this method needs a lot of vector registers, and the knots as well as the weights depend on the position of the singular point so that the CPU–time for the outer integration increases drastically with increasing numbers of knots. Polar coordinates or triangle coordinates yield better accuracies than all the methods mentioned before. The only disadvantage is that we need a lot of knots for the integration over the angular coordinate, especially if the singular points are very near by the boundary of the domain of integration. But this disadvantage can be eliminated by using polar coordinates and additionally the angular transformation of Hayami & Brebbia [18]. Thus we are able to evaluate the inner integral with an accuracy of 10^{-4} using only 16 knots whereas all the other methods produce cubature errors one to three orders larger.

We use Gauss–Legendre product formulas for the outer integration, but the accuracy we get for the complete, weakly singular integral (15) is very poor. The reasons for that are the discontinuity of the function ϑ for $\rho = 0$ and the structure of the function ϑ which requires some taylor expansions to compute this function numerically stable. Moreover, the accuracy decreases with increasing roughness of the topography. That is why we get only relative accuracies of the order 10^{-2} – 10^{-3} with reasonable numbers of knots. A further problem is the balancing of the numbers of knots for the outer and inner integration. We get the best results making the same choice.

Table 4 shows the three parts $a_{ii,r}$, $a_{ii,c}$ and $a_{ii,s}$ for some boundary elements Γ_i with different roughness of topography. We see that the regular parts are considerably larger than the weakly singular and strongly singular parts, the order depends on the roughness of the topography. Moreover, we observe a neutralization between the weakly singular and the strongly singular parts which is the larger the flatter the topography. That means, we primarily have to compute the regular parts $a_{ii,r}$ with high degree of precision whereas larger errors are allowed for the weakly singular and strongly singular parts of the main diagonal elements.

Table 4 Main diagonal coefficient a_{ii} for different roughness of topography 69 (flat), 1 (moderate) 1027 (rough)

	parts of the main diagonal coefficient a_{ii}			
i	$a_{ii,r}$	$a_{ii,c}$	$a_{ii,s}$	$a_{ii,c} + a_{ii,s}$
69	25.133713	–0.003441	0.004372	+0.000930
1	25.144520	–0.013708	0.014115	+0.000407
1027	25.129367	–0.154737	0.132953	–0.021784

The off–diagonal terms are defined by regular double integrals over parts of the earth's surface. In the case where the outer and inner regions of integration are not adjacent the kernels are well behaved and we can use simple Gauss–Legendre product formulas for the efficient computation of these integrals. Most of the elements can be computed with relative accuracy of 10^{-3} – 10^{-4} using only 16 knots. But the main problem is their large number: If we discretize the boundary surface Γ using n boundary elements, we have to compute approximately n^2 of these elements, i.e. for $n = 10000$ approximately 10^8 double integrals over parts of the earth's surface. To save CPU–time we have to choose the number of knots in subordination to the distance between outer and inner domain of integration. But this means logical IF–statements in FORTRAN source codes which can drastically reduce the performance of the algorithm so that CPU–time increases. This loss of speed can only be limited if we use special vector directives which are machine dependent, e.g. the so called VOCL–instructions on Siemens/Fujitsu VP–pipeline machines.

In the case where outer and inner domains of integration are adjacent the kernel of the integral shows a behaviour similar to those of singular integrals because it is proportional to $|x-y|^{-2}$ so that function values drastically increase for small distances between outer (x) and inner (y) knots. We examine several methods which can deal with such integrals, like the cubic parameter transformation proposed by Telles [4] applied to the cartesian coordinates, polar coordinates with the outer knots as poles with and without an additional cubic parameter transformation of Telles applied to the polar distance. We exclude methods based on a detailed subdivision of the domain of integration [see e.g. 10, 19] because this subdivision depends on the position of the point x and is for that reason very time–consuming especially on vector pipeline machines. Beyond that the accuracy is in relation to the number of knots very poor in comparison with other methods mentioned before. In table 5 we find some results for the evaluation of the <u>inner</u> integral for different positions of the point x relative to the domain of integration. From all the methods the cubic parameter transformation of Telles applied to the cartesian coordinates {2} and the use of polar coordinates combined with Gauss–Legendre product formulas {3} yield the best results. For small numbers of knots the

Table 5 Evaluation of quasi–singular integrals with different cubatures.
<u>Here:</u> Relative cubature errors for the inner integral.
Domain of integration: [–1,1]•[–1,1]. N denotes the number of knots.

	$x=(-1.1\ ,\ 0.0)$ CPU–times in milliseconds							
N	{1}	CPU	{2}	CPU	{3}	CPU	{4}	CPU
9	3.3 (–1)	9.9	2.4 (–1)	10.4	1.9 (–1)	11.1	1.6 (–1)	11.7
36	9.4 (–1)	13.4	2.7 (–1)	15.0	4.7 (–3)	16.8	6.7 (–3)	18.8
81	7.0 (–1)	22.3	1.4 (–2)	26.0	1.9 (–3)	29.9	1.7 (–3)	34.1
144	3.0 (–1)	40.3	6.5 (–3)	46.8	9.4 (–4)	53.7	8.5 (–4)	61.0
441	4.2 (–2)	79.9	4.3 (–5)	99.2	3.3 (–5)	119.8	3.3 (–5)	141.
	$x=(-1.005\ ,\ 0.0)$ CPU–times as above							
N	{1}	CPU	{2}	CPU	{3}	CPU	{4}	CPU
9	8.5 (–1)		1.4 (±0)		8.0 (–1)		5.9 (–2)	
36	7.1 (–1)		6.7 (–1)		4.6 (–1)		2.6 (–3)	
81	2.9 (±0)		4.4 (–1)		2.4 (–1)		7.1 (–4)	
144	1.1 (±0)		1.1 (–1)		1.3 (–1)		3.0 (–5)	
441	3.9 (±0)		3.0 (–2)		2.0 (–2)		1.2 (–4)	

combination of polar coordinates with a cubic transformation applied to the polar distance {4} is superiour to all other methods but the increase in accuracy with increasing numbers of knots is considerably smaller as for the cubic transformation of Telles applied to the cartesian coordinates. The use of simple Gauss–Legendre product formulas {1} yields insufficient accuracies. A disadvantage of all methods using polar coordinates is the required determination of the position of the outer knot relative to the inner domain of integration to determine the lower and upper integration bounds for the inner integration. This process requires additional CPU–time. All in all the cubic parameter transformation of Telles applied to the cartesian coordinates and combined with Gauss–Legendre product formulas is the best compromise between accuracy and CPU–time required. Its disadvantage is the need of main memory because the knots and weights for the inner integration depend on the position of the knots for the outer integration, e.g. to each outer knot belongs a special cubature formula. For the outer integration we use Gauss–Legendre product formulas and get the smallest cubature errors for choosing the same number of knots for inner and outer integration.

Table 6 shows some characteristics for computing all the elements of the linear system of equations. The characteristics are based on a discretization of the boundary surface into 10000 boundary elements leading to a dimension 10000·10000 for the equation matrix. The computation of all the elements takes approximately 70 CPU–minutes on a vector pipeline machine Siemens/Fujitsu VP 400–EX. Approximately 97 % of these fall to the share of the off–diagonal elements whereas the computation of the main diagonal takes approximately 1.9% and that of the inhomogeneous part only 1% of the total CPU–time.

Table 6 Some characteristics for evaluating the coefficients of the equation matrix

type of coefficient	number of knots for 1 coefficient (inner/outer Integration/total)	total number of coefficients/knots	timing (seconds) CPU–time	I/O–time
main diagonal				
regular	—/400/400	$10000/4\cdot10^6$	9.4	5.4
strongly singular	—/1600/1600	$10000/16\cdot10^6$	50.3	2.6
weakly singular	36/36/1296	$10000/13\cdot10^6$	13.6	2.4
off–diagonal				
quasi–singular	64/64/4096	$78804/3.2\cdot10^8$	919.7	17.5
regular	25/25/625	$99911196/6.2\cdot10^{10}$	2800.8	1750.0

SOLUTION OF THE LINEAR SYSTEM OF EQUATIONS

It is well known that boundary element methods lead to dense linear systems. Moreover, using the Galerkin–Bubnov discretization the equation matrix is only symmetric if the integral equation is self–adjoint. In our case this means that we have to solve a dense unsymmetric linear system of dimension 10000·10000. Using data type DOUBLE PRECISION we need approximately 800 MBytes of memory to store the coefficient matrix.

As the main storage of the vector pipeline computer Siemens/Fujitsu VP–400 EX is limited to 256 MBytes MSU plus 256 MBytes VSU, iterative solution methods are out of the question because a lot of I/O–operations are necessary and this operations are very time–consuming. Instead of iterative methods we use a direct equation solver based on a LU–factorization using level 2 and 3 Basic Linear Algebra Subprograms (BLAS), optimized for the Siemens/Fujitsu Vector Processor (VP) family and columnwise pivoting [see 20 for details].

The solution of the linear system $A\cdot\boldsymbol{\alpha} = \boldsymbol{h}$ is split up into two steps: First we factorize the matrix A into a product $P\cdot L\cdot U$ where P denotes the permutation matrix and L (U) the lower (upper) triangular matrix. Next we solve the linear system by forward and backward elimination i.e. $\boldsymbol{\gamma} = P^{-1}\cdot\boldsymbol{h}$, $\boldsymbol{\beta} = L^{-1}\cdot\boldsymbol{\gamma}$ and $\boldsymbol{\alpha} = U^{-1}\cdot\boldsymbol{\beta}$. For the LU–factorization as the most time–consuming step we use the so called "left–looking–method" [21, 22]: Each of the matrizes L and U is computed columnwise where in each step we replace one column of the original matrix by one column of L and one column of U. This method has been proved to be the fastest method on the pipelined machine VP–400 EX. The critical points of the algorithm are the I/O–perations because the I/O–time is much longer than the time for the factorization. We reduce the I/O–time using certain runtime options and asynchronous I/O so that I/O–operations and CPU–usage are done in parallel.

The whole matrix A of dimension N·N is split up into submatrizes A_i, i=1,...,m. The submatrix A_i, i=1,...,m–1 consists of the columns (i–1)·NB+1 to i·NB and the submatrix A_m consists of the columns (N–1)·NB+1 to N. The choice of NB depends on the available main storage and the dimension N of the linear system. We denote by A(j,k) the submatrix which consists of the rows (k–1)·NB+1 to j·NB of the original matrix A and by A^k the first min[(k–1)·NB,N] columns of A. We use the following algorithm [20]:

```
start reading A_1
for j=1,m
        start reading L_1
        wait until U_{j-1} and L_{j-1} have been written
        wait until A_j has been read
        for i=1,j-1
                apply row interchanges of i-th block to A_j
                wait until L_i has been read
                start reading L_{i+1}
                U_j(i,i) = L_i^{-1}(i,i)·A_j(i,i)
                A_j(m,i+1) = A_j(m,i+1) - L_i(m,i+1)·U_j(i,i)
        end for i
        factorize A_j(m,j): A_j(m,j) = P_j·L_j(m,j)·U_j(j,j)
        start writing L_j and U_j and reading A_{j+1}
end for j
```

For the factorization of the submatrix $A_j(m,j)$ we use the modified "right–looking method" [22]. Table 7 contains some characteristics for the solution of several linear systems. Accordingly it takes 632 seconds CPU–time (including 626 seconds of vector unit (VU) time) to solve a dense unsymmetric linear system of dimension 10000·10000 with 6 right sides on the pipelined machine VP–400 EX. The execution time was 4020 seconds most of it caused by I/O–operations. Using the advanced pipelined machine Siemens/Fujitsu S400/10 the CPU–time is reduced to 425 seconds (including 416 seconds vector unit (VU) time) and the execution time to 2670 seconds. Each run requires 420 MBytes main storage and the choice for the parameter NB was 910 so that the equation matrix A was split up into 11 blocks. With reducing dimension of the linear system both the CPU–time and the execution time drastically decrease so that e.g. the solution of a 3600·3600 linear system takes only 35 CPU–seconds and 178 seconds execution time. The next generation of pipelined machines, which will be installed at the computer centre of the university of Karlsruhe in 1991, promises CPU—times of approximately 2–3 minutes for the solution of a 10000·10000 linear system and a drastically reduced execution time because an increase in main memory of more than 1 GBytes.

Table 7 Performance for the solution of some linear systems of equations using vector pipeline machines. NRS: Number of right sides.

Dimension	NRS	NB	Machine	MFLOPS	timing (seconds)		
					CPU	VU	Exec.
10000	6	910	400–EX	1057	632	626	4020
10000	6	910	400/10	1570	425	416	2670
8100	8	1100	400–EX	1011	352	348	1650
6400	8	1400	400–EX	990	177	175	1270
3600	8	2500	400–EX	906	35	34	178
2500	8	2500	400–EX	922	11.4	11.0	66
400	8	400	400–EX	275	0.16	0.09	2.08

COMPUTATION OF THE GRAVITATIONAL POTENTIAL

In opposition to finite element methods, boundary element methods do not give us directly the solution of the original partial differential equation in its region of definition. Rather we have to compute it point by point from its values on the boundary (in the case of direct formulation) or from the values of a density on the boundary surface (in the case of indirect formulation). Using the potential of a single layer we have to compute numerically a surface integral to get the gravitational potential in a point in the outer space of the earth or on the earth's surface:

$$V(\mathbf{x}) = V_0(\mathbf{x}) + \frac{1}{4\cdot\pi}\cdot\int\limits_{\mathbf{y}\in\Gamma} \frac{f(\mathbf{y})}{|\mathbf{x}-\mathbf{y}|}\, d\Gamma(\mathbf{y}) \quad , \mathbf{x}\in\Omega_a\cup\Gamma. \tag{20}$$

In the case where the surface density f is approximated using piecewise polynomial elements and the boundary is represented by flat triangular elements we find analytical solutions of the integral (20) e.g. in [23, 24]. For an approximation of the boundary with higher order elements we have to evaluate the integral (20) numerically. If point $\mathbf{x}$ is situated on the surface, the integral is weakly singular and we can use the same cubatures which have been discussed before in connexion with the evaluation of the elements of the Galerkin–equations. If the distance $\mathrm{dist}(\mathbf{x},\Gamma)$ is greater than a certain number λ depending on the maximum diameter h of the boundary elements we can use simple standard cubatures, especially Gauss–Legendre product formulas. As their degree of precision must be tuned to the smoothness of our integrand we have to use in general composed cubature formulas. For distances $\mathrm{dist}(\mathbf{x},\Gamma) < \lambda$ we must be careful because the integrand shows a behaviour similar to that of singular integrals and using standard cubatures results in large cubature errors. The reason for that is the term $|\mathbf{x}-\mathbf{y}|^{-1}$ which causes large variations of the integrand if $\mathrm{dist}(\mathbf{x},\Gamma)$ becomes smaller and smaller. Similar problems arise if we want to evaluate e.g. the gradient of the gravitational potential in points near by the earth's surface where the integrand behaves like $|\mathbf{x}-\mathbf{y}|^{-2}$. In both cases we have to develop special cubature formulas which can deal with such strong variations.

We examine several cubature formulas, e.g. the angular and radial transformation proposed by Hayami & Brebbia [18] based on the orthogonal projection of the boundary element into the tangential plane of the orthogonal projection $\mathbf{x}_f$ of $\mathbf{x}$ onto the boundary surface {2} and the cubic parameter transformation proposed by Telles [4] with {3} and without {4} parameter optimization and compare them with simple Gauss–Legendre product formulas {5}. Besides we use a modified version of the angular and radial transformation {1}. The modification avoids the approximation of the region of integration used by Hayami & Brebbia which can cause large errors if the topography is rough. Moreover, the knots which are primarily defined in the tangential plane must be map onto the boundary element Γ_i which can be done only iteratively and takes some CPU–time. These disadvantages can be avoided if there is a parameter representation of the boundary Γ available so that we are able to replace the tangential plane by the parameter domain. Using (10) we get

$$\int\limits_{\mathbf{y}\in\Gamma_i} \frac{f(\mathbf{y})}{|\mathbf{x}-\mathbf{y}|}\, d\Gamma(\mathbf{y}) = \int\limits_{\eta\in S_i} \frac{f\circ\varphi(\eta)}{|\mathbf{x}-\mathbf{y}|}\cdot\sqrt{g(\eta)}\, d\eta =: \int\limits_{\eta\in S_i} \frac{h(\eta)}{|\mathbf{x}-\varphi(\eta)|}\, d\eta \quad , h(\eta) := f\circ\varphi(\eta)\cdot\sqrt{g(\eta)} \tag{21}$$

and after introducing polar coordinates (ρ,ψ) with regard to $\xi = \varphi^{-1}(\mathbf{x}_f)$

$$\int_{y\in\Gamma_i} \frac{f(y)}{|x-y|}\, d\Gamma(y) = \sum_{k=1}^{4} \int_{\psi_k}^{\psi_{k+1}} \int_{0}^{R_k(\psi)} \frac{h(\eta)}{|x-\varphi(\eta)|}\cdot \rho\, d\rho d\psi(24) \quad . \tag{22}$$

After using the angular transformation

$$t(\psi) = \frac{1}{2}\cdot d_i \cdot \ln\left[\frac{1+\sin[\psi-k\cdot\frac{\pi}{2}]}{1-\sin[\psi-k\cdot\frac{\pi}{2}]}\right] \quad , \psi(t) = k\cdot\frac{\pi}{2} + \arcsin\left[\frac{\exp(\frac{2}{d_k}\cdot t-1)}{\exp(\frac{2}{d_k}\cdot t+1)}\right] \tag{23}$$

and the radial transformation

$$\rho d\rho = (\rho^2+r_f^2)\, ds \Longrightarrow s = \frac{1}{2}\cdot \ln(\rho^2 + r_f^2) \text{ resp. } \rho = (\exp(2s)-r_f^2)^{1/2} \tag{24}$$

(the integration constant has been choosen so that $s(0) = \ln(r_f)$) we get finally for the integral over the triangle Δ_{ik}, k=1,...,4:

$$\int_{\psi_k}^{\psi_{k+1}} \int_{0}^{R_k(\psi)} \frac{h(\eta)}{|x-\varphi(\eta)|}\cdot\rho\, d\rho d\psi = \int_{t_k(\psi_k)}^{t_k(\psi_{k+1})} \int_{s(0)}^{s\circ R_k\circ\psi(t)} \frac{h}{r}\cdot(\rho^2+r_f^2)\cdot[R_k\circ\psi(t)]^{-1}\, ds dt \tag{25}$$

where $r := |x-\varphi(\eta)|$. $R_k(\psi)$ denotes the equation of the boundary k of the rectangular element in polar coordinates and d_k the distance of ξ from this boundary. Table 8 contains for different values of $r_f = |x-x_f| = \text{dist}(x,\Gamma)$ and different numbers of knots the results using the cubature methods mentioned above.

Table 8 Numerical evaluation of the potential $v(x)$. Size of the boundary element: $1000\cdot 1000$ m². 64 knots/boundary element

dist(x,Γ) (meter)	cubature method {1}	{2}	{3}	{4}	{5}
10000	8.5 (−10)	1.4 (−6)	1.0 (−9)	9.3 (−10)	1.5 (−10)
1000	2.1 (−7)	4.5 (−6)	2.1 (−8)	9.4 (−9)	2.1 (−8)
100	5.5 (−6)	1.3 (−5)	5.4 (−6)	2.0 (−4)	6.5 (−4)
10	9.7 (−6)	1.8 (−5)	3.0 (−4)	1.4 (−4)	7.2 (−3)
1	5.9 (−5)	7.0 (−5)	3.4 (−4)	3.6 (−3)	8.7 (−3)
0.1	3.3 (−4)	3.5 (−4)	5.2 (−4)	3.7 (−3)	8.9 (−3)
0.01	1.1 (−3)	1.2 (−3)	5.2 (−4)	3.7 (−3)	8.9 (−3)

REFERENCES

[1] KLEES, R.: Solution of the fixed geodetic boundary–value problem using the boundary element method. Thesis to be published in 1991 (in german).

[2] ZEIDLER, E.: Nonlinear functional analysis and its applications. Vol. I: Fixed–Point Theorems, Springer, New York 1986.

[3] HIBBS, T.T.: C¹–continuous representations and advanced singular kernel integrations in the three–dimensional boundary integral method. Thesis, Department of Mathematics & Statistics, Teesside Polytechnic, 1989.

[4] TELLES, J.C.F.: A self–adaptive co–ordinate transformation for efficient numerical evaluation of general boundary element integrals. Int. J. Numer. Methods Engrg. 24 (1987), pp. 959–973.

[5] MUROTA, K., IRI, M.: Parameter tuning and repeated application of the IMT–type transformation in numerical quadrature. Numer. Math. 38 (1982), pp. 347–363.

[6] MORI, M.: Quadrature formulas obtained by variable transformation and the DE–rule. J. Comput. and Appl. Math. 12–13 (1985), pp. 119–130.

[7] JENG, G., WEXLER, A.: Isoparametric finite element variational solution of integral equations for 3D–fields. Int. J. Numer. Methods Engrg. 11 (1977), pp. 1455–1477.

[8] CRISTESCU, M., LOUBIGNAC, G.: Gaussian quadrature formulas for functions with singularities in r^{-1} over triangles and quadrangles. In: Brebbia, C.A. (Ed.): Recent advances in boundary element methods. Pentech Press, London 1987.

[9] PINA, H.L.G., FERNANDES, J.L.M., BREBBIA, C.A.: Some numerical integration formulae over triangles and squares with a R^{-1} singularity. Appl. Math. Mod. 5 (1981), pp. 209–211.

[10] LACHAT, J.C., WATSON, J.O.: Effective numerical treatment of boundary integral equations: A formulation for three–dimensional elastostatics. Int. J. Numer. Methods Engrg. 10 (1976), pp. 991–1005.

[11] LEAN, M.H., WEXLER, A.: Accurate numerical integration of singular boundary element kernels over boundaries with curvature. Int. J. Numer. Methods Engrg. 21 (1985), pp. 211–228.

[12] ALIABADI, M.H., HALL, W.S., PHEMISTER, T.G.: Taylor Expansions for singular kernels in the boundary element method. Int. J. Numer. Methods Engrg. 21 (1985), pp. 2221–2236.

[13] ALIABADI, M.H., HALL, W.S.: Weighted gaussian methods for three–dimensional boundary element kernel integration. Comm. Appl. Numer. Methods 3 (1987), pp. 89–96.

[14] LI, H.–B., HAN, G.–M., MANG, H.A.: A new method for evaluating singular integrals in stress analysis of solids by the direct boundary element method. Int. J. Numer. Methods Engrg. 21 (1985), pp. 2071–2098.

[15] ALIABADI, M.H., HALL, W.S.: The regularising transformation integration method for boundary element kernels. Comparison with series expansion and weighted gaussian integration methods. Engrg. Anal. 5 (1988).

[16] Ying, Lung–An: Some special interpolation formulae for triangular and quadrilateral elements. Int. J. Numer. Methods Engrg. 18 (1982), pp. 959–966.

[17] ROBINSON, I., DE DONCKER, E.: Automatic computation of improper integrals over a bounded or unbounded planar region. Computing 16 (1981), pp. 253–284.

[18] HAYAMI, K., BREBBIA, C.A.: Quadrature methods for singular and nearly singular integrals in 3–D boundary element method. In: Brebbia, C.A. (Ed.): Boundary Elements X, Proc. 10th Int. Conf., Southampton, U.K., Vol. 1, pp. 237–264.

[19] JUN, L., BEER, G., MEEK, J.L.: Efficient evaluation of integrals of order r^{-1}, r^{-2}, r^{-3} using gauss quadrature. Engrg. Anal. 2 (1985), pp. 118–123.

[20] GEERS, K.: Solving large dense linear systems on Siemens/Fujitsu VP. Internal Report, University of Karlsruhe, Computer Center, 1990.

[21] ANDERSON, E., DONGARRA, J.: Evaluating block algorithm variants in LAPACK. LAPACK working note 19, CS–90–103. Paper submitted to the proceedings of the fourth SIAM conference on parallel processing for scientific computing, held in Chicago, Illinois, 1989.

[22] GEERS, K.: Optimization of level 2 BLAS for Siemens VP Systems. Report No. 37.89, University of Karlsruhe, Computer Center, 1989.

[23] BOSCH, W.: Geschlossene Integration von Potentialkernen beim Modell der einfachen Schicht mit stückweise ebenem Rand. Mitteilungen aus dem Institut für Theoretische Geodäsie der Universität Bonn Nr. 49, Bonn 1977.

[24] HACKBUSCH, W.: Integralgleichungen. Teubner–Verlag, Stuttgart 1989.

The analytical integration of boundary integrals for plate bending

B. Knöpke
Gh Kassel, FB 14, Lehrstuhl für Baustatik
Mönchebergstr. 7, D-3500 Kassel, Germany

Summary

Serious problems arise when Gaussian quadrature is employed for the integrations required in the direct boundary element method (DBEM). You will run in a lot of trouble if you are going to compute domain points near the boundary. These problems can be overcome if you carry out the integration analytically. The technique is developped in this paper.

Introduction

One of the major problems in the application of the direct boundary element method (DBEM) to elliptic partial differential equations of fourth order are the many hypersingular kernels within. This distinguishes fourth order equations from second order equations. In second order theory you can avoid these kernels altogether if you use an appropriate formulation.

In the case of the plate equation the integral equation for the slope $\frac{\partial w}{\partial n}$ already contains a hypersingular kernel (r^{-2}). For the calculation of the internal actions (bending moment and shear force) you need to know the second and the third derivatives of the deflection w. In the representation formulas for these terms appear even stronger singular kernels (r^{-3} and r^{-4}) respectively.

The program BE-PLATTE, a professional program for the analysis and design of concrete and steel plates, is based on the direct boundary element method and solves the integral equations of the deflection w and the slope $\frac{\partial w}{\partial n}$ by a collocation method. The program is restricted to straight elements.

The design of the reinforcement within a plate structure depends on the distribution of the bending moments in the interior of the plate. For that purpose the program lays after a choice of a mesh width by the user a mesh of stress points over the plate and calculates in this points the bending moments. Intuitively you would expect that a regular mesh is chosen. Unfortunately the generated mesh which depends on the shape of the plate is in some cases very irregular. This leads to illegible reinforcement layouts, s. Fig. 1

The reasons are to be found in the restrictions which the program has to obey in generating the mesh. These restrictions are due to the fact that the program renders

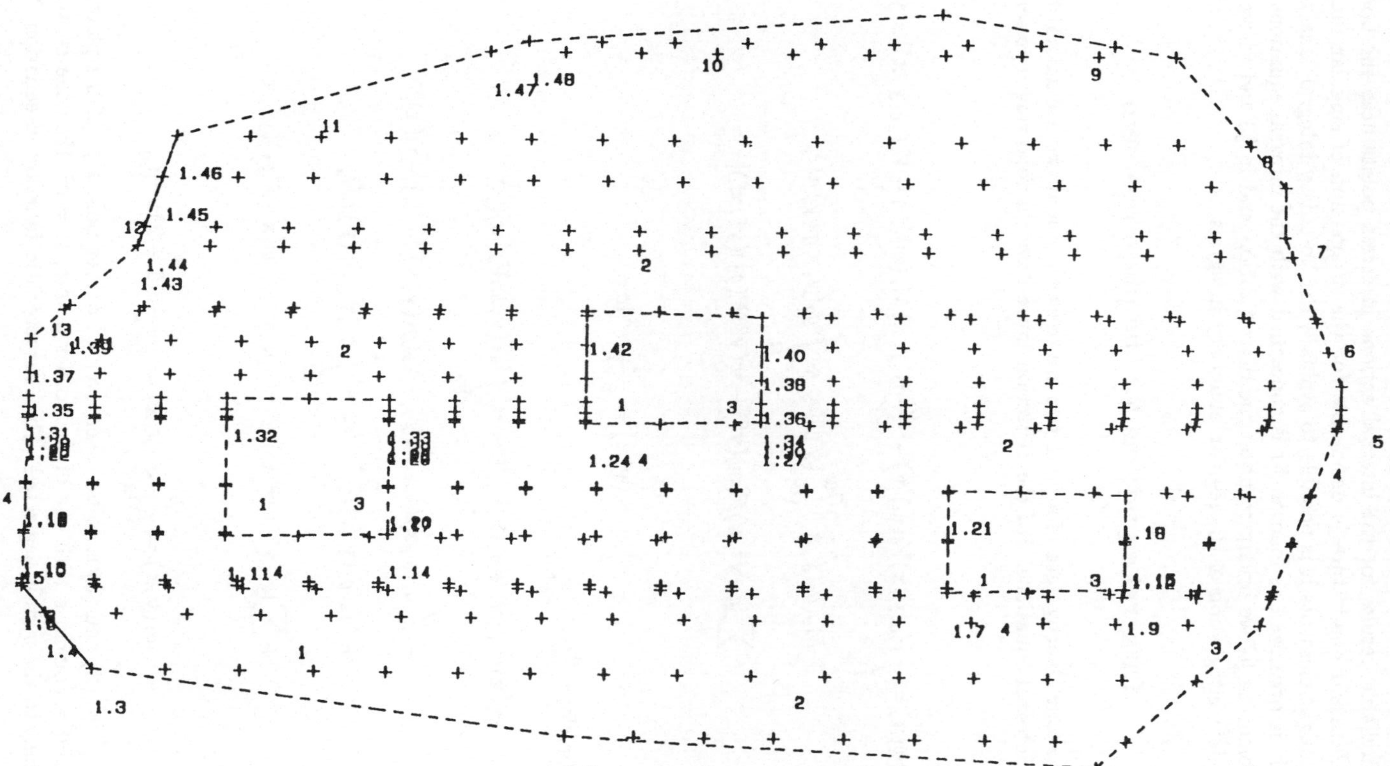

Fig.1 layout of reinforcement

totally unreliable results for the internal actions in stress points near the boundary because Gaussian quadrature is unsuitable for the integration of singular functions. But for straight elements it is possible to evaluate the boundary integrals analytically. The article is organized as follows: §1 is concerned with the integral equations of the plate problem, in §2 we briefly mention the discretization and in §3 and §4 we dicuss the analytical integration of the regular boundary integrals.

Integral equations for the plate problem

Of the four Cauchy-data of a plate two are prescribed and two are unknown so we need two integral equations. For the deflection w we have the following representation formula

$$
\begin{aligned}
c(\mathbf{x})w(\mathbf{x}) = \int_{\Gamma} & [g_0(\mathbf{y},\mathbf{x})V_\nu(w)(\mathbf{y}) - \frac{\partial}{\partial \nu} g_0(\mathbf{y},\mathbf{x})M_\nu(w)(\mathbf{y}) - V_\nu(g_0(\mathbf{y},\mathbf{x}))w(\mathbf{y}) \\
& + M_\nu(g_0(\mathbf{y},\mathbf{x}))\frac{\partial w}{\partial \nu}(\mathbf{y})]\, ds_{\mathbf{y}} + \int_{\Omega} g_0(\mathbf{y},\mathbf{x})p(\mathbf{y})\, d\Omega_{\mathbf{y}} \\
& + \sum_c [g_0(\mathbf{y}^c,\mathbf{x})F(w)(\mathbf{y}^c) - w(\mathbf{y}^c)F(g_0)(\mathbf{y}^c,\mathbf{x}^c)]
\end{aligned} \tag{2}
$$

where

$$
g_0(\mathbf{y},\mathbf{x}) = \frac{1}{8\pi K} r^2 \ln r
$$

and for the slope $\frac{\partial w}{\partial n}$

$$
\begin{aligned}
c_1(\mathbf{x})w_{,1}(\mathbf{x}) + c_2(\mathbf{x})w_{,2}(\mathbf{x}) = \int_{\Gamma} & [g_1(\mathbf{y},\mathbf{x})V_\nu(w)(\mathbf{y}) \\
& - \frac{\partial}{\partial \nu} g_1(\mathbf{y},\mathbf{x})M_\nu(w)(\mathbf{y}) - V_\nu(g_1(\mathbf{y},\mathbf{x}))[w(\mathbf{y}) - w(\mathbf{x})] \\
& + M_\nu(g_1(\mathbf{y},\mathbf{x}))\frac{\partial w}{\partial \nu}]\, ds_{\mathbf{y}} + \int_{\Omega} g_1(\mathbf{y},\mathbf{x})p(\mathbf{y})\, d\Omega_{\mathbf{y}} \\
& + \sum_c [g_1(\mathbf{y}^c,\mathbf{x})F(w)(\mathbf{y}^c) - [w(\mathbf{y}^c) - w(\mathbf{x})]F(g_1)(\mathbf{y}^c,\mathbf{x})]
\end{aligned} \tag{3}
$$

where

$$
g_1(\mathbf{y},\mathbf{x}) = \frac{\partial}{\partial n_{\mathbf{x}}} g_0(\mathbf{y},\mathbf{x}) = \frac{1}{8\pi K} r(1 + 2\ln r)r_{,n}.
$$

The sum $\sum$ in (2) and (3) is to be taken over all corner points $\mathbf{y}^c$ of the plate. If the source point $\mathbf{x}$ coincides with one of the corner points, $\mathbf{x} = \mathbf{y}^c$, then the contribution of this point to the sum is neglected. The characteristic function in equation (2) is

$$
c(\mathbf{x}) = \begin{cases} 1, & \mathbf{x} \in \Omega, \\ \Delta\varphi/2\pi, & \mathbf{x} \in \Gamma, \\ 0, & \mathbf{x} \in \Omega^c = \mathbf{R}^2 \setminus \overline{\Omega}. \end{cases}
$$

And the characteristic functions c_1 and c_2 in equation (3) are defined as follows:

$$c_1(\mathbf{x}) = \begin{cases} n_1, \\ \dot{c}_1(\mathbf{x}), \\ 0, \end{cases} \qquad c_2(\mathbf{x}) = \begin{cases} n_2, & \mathbf{x} \in \Omega, \\ \dot{c}_2(\mathbf{x}), & \mathbf{x} \in \Gamma, \\ 0, & \mathbf{x} \in \Omega^c. \end{cases}$$

Their boundary values

$$\dot{c}_1(\mathbf{x}) = \frac{\Delta\varphi}{2\pi} n_1 + \frac{\nu}{2\pi}[\frac{1}{2}\sin 2\varphi\, n_1 + \sin^2\varphi\, n_2]_{\varphi_2}^{\varphi_1}, \tag{4}$$

$$\dot{c}_2(\mathbf{x}) = \frac{\Delta\varphi}{2\pi} n_2 + \frac{\nu}{2\pi}[\sin^2\varphi\, n_1 - \frac{1}{2}\sin 2\varphi\, n_2]_{\varphi_2}^{\varphi_1} \tag{5}$$

depend on the angles φ_1 und φ_2 which the two tangents at the boundary point form with the x_1-axis. At smooth points, where $\Delta\varphi = \varphi_1 - \varphi_2$ is $180° \hat{=} \pi$ the brackets in Equations (4) and (5) vanish,

$$[\sin 2\varphi]_{\varphi_2}^{\varphi_1} = \sin 2\varphi_1 - \sin 2\varphi_2 = 0,$$
$$[\sin^2\varphi]_{\varphi_2}^{\varphi_1} = \sin^2\varphi_1 - \sin^2\varphi_2 = 0$$

so that at such points

$$\dot{c}_1(\mathbf{x}) w_{,1}(\mathbf{x}) + \dot{c}_2(\mathbf{x}) w_{,2}(\mathbf{x}) = \frac{1}{2}\frac{\partial w}{\partial n}(\mathbf{x}).$$

If we place the point $\mathbf{x}$ on the boundary then Equations (2) and (3) constitute two integral equations for the four boundary values

$$w, \quad \frac{\partial w}{\partial n}, \quad M_n(w), \quad V_n(w)$$

and the corner terms

$$w(\mathbf{x}^c), \quad F(\mathbf{x}^c)$$

of a Kirchhoff plate. As two of the four boundary functions and one of the two conjugated corner terms are determined by support and loading conditions the two integral equations can be solved for the two unknown functions.

The discretization of the integral equations

For the approximate solution of the two integral equations (2) and (3) we use point collocation. For that we choose K points $\mathbf{x}^k$ on the boundary, the collocation points. We interpolate the boundary functions by polynomials which are defined piecewise over subregions of the boundary called boundary elements.

To determine the unknown nodal values of the piecewise polynomial functions, you don't demand any more that both integral equations are fulfilled for all $\mathbf{x} \in \Gamma$, but for the collocation points $\mathbf{x}^k$, $k = 1, \cdots, K$.

Now for the boundary integrals to exist, the piecewise polynomial functions must be sufficiently smooth at the collocation points. The approximation of the deflection must be C^1 at the node (the tangential derivative must be continuous) and the approximation of the slope must be C^0, i. e. continuous.

Therefore we approximate the deflection w along the boundary elements Γ_e by Hermite-polynomials and the other boundary functions by piecewise linear polynomials.

Calculation of the boundary integrals

The values of the Cauchy principal values in the case of straight elements are given in [1]

Now we turn to the regular boundary integrals. Over straight elements the product of kernel functions times boundary layer can be integrated easily analytically if we switch to a system of coordinates s, d That is centered at the source point $\mathbf{x}$ and whose axis are parallel (s) and orthogonal (d) respectively to the element and its normal vector, s. Fig. 2.

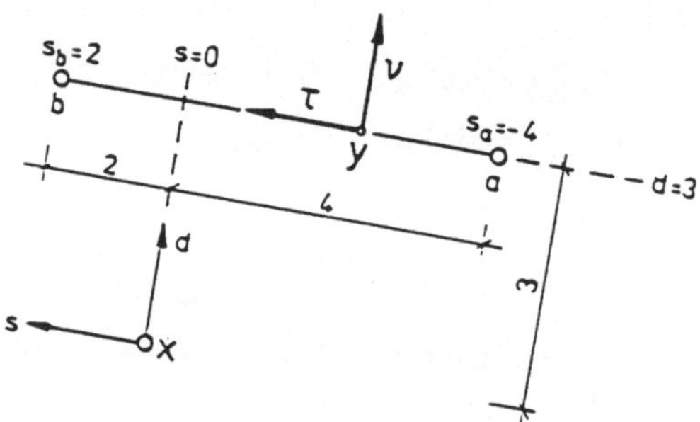

Fig.2 d-s-coordinate system

In that case the distance

$$r = [d^2 + s^2]^{\frac{1}{2}},$$

varies only with s (which is up to a constant 'translational' term the arc-length s on the element) and the normal and tangential derivative of the distance r become

$$r_\nu = \frac{d}{r}, \quad r_\tau = \frac{s}{r}, \quad d = \text{fixed},$$

so that the kernel functions can be written as

$$\begin{aligned}
g_0(\mathbf{y}, \mathbf{x}) &= \frac{1}{8\pi K} r^2 \ln r, \\
\frac{\partial}{\partial \nu} g_0(\mathbf{y}, \mathbf{x}) &= \frac{1}{8\pi K} d(1 + 2\ln r), \\
M_\nu(g_0(\mathbf{y}, \mathbf{x})) &= -\frac{1}{8\pi}[(1+3\nu) + 2(1+\nu)\ln r + 2(1-\nu)\frac{d^2}{r^2}], \\
V_\nu(g_0(\mathbf{y}, \mathbf{x})) &= -\frac{1}{8\pi}[2(1+\nu)\frac{d}{r^2} + 4(1-\nu)\frac{d^3}{r^4}]
\end{aligned}$$

and

$$
\begin{aligned}
g_1(\mathbf{y},\mathbf{x}) =& -\mathbf{n}\cdot\nu\frac{1}{8\pi K}d(1+2\ln r)-\mathbf{n}\cdot\tau\frac{1}{8\pi K}s(1+2\ln r),\\
\frac{\partial}{\partial\nu}g_1(\mathbf{y},\mathbf{x}) =& -\mathbf{n}\cdot\nu\frac{1}{8\pi K}[1+2\ln r+2\frac{d^2}{r^2}]-\mathbf{n}\cdot\tau\frac{1}{4\pi K}\frac{ds}{r^2},\\
M_\nu(g_1(\mathbf{y},\mathbf{x})) =& \mathbf{n}\cdot\nu\frac{1}{4\pi}[(1+\nu)\frac{d}{r^2}+2(1-\nu)\frac{ds^2}{r^4}]\\
&+\mathbf{n}\cdot\tau\frac{1}{4\pi}[(1+\nu)\frac{s}{r^2}-2(1-\nu)\frac{d^2s}{r^4}],\\
V_\nu(g_1(\mathbf{y},\mathbf{x})) =& \mathbf{n}\cdot\nu\frac{1}{4\pi}[4(2\nu-1)\frac{d^2}{r^4}+8(1-\nu)\frac{d^4}{r^6}-(1+\nu)\frac{1}{r^2}]\\
&+\mathbf{n}\cdot\tau\frac{1}{4\pi}[2(1+\nu)\frac{ds}{r^4}+8(1-\nu)\frac{d^3s}{r^6}].
\end{aligned}
$$

The polynomial basis functions defined on an element Γ_i can be expressed in terms of s as well. If we multiply these boundary layers with the kernel functions then we see that all the boundary integrals can be expressed in terms of the following five integrals

$$
\begin{aligned}
I^k &= \int s^k \ln r\, ds = \frac{1}{k+1}[s^{k+1}\ln r - J^{k+2}],\\
J^k &= \int \frac{s^k}{r^2}\, ds = \begin{cases} [\frac{s^{k-1}}{k-1} - d^2 J^{k-2}], & k\ge 2,\\ \ln r, & k=1,\\ \frac{1}{d}\arctan\frac{s}{d}, & k=0,\end{cases}\\
K^k &= \int \frac{ds^k}{r^4}\, ds = \begin{cases} \frac{d}{2}[-\frac{s^{k-1}}{r^2}+(k-1)J^{k-2}], & k\ge 2,\\ \frac{d}{2r^2}, & k=1,\\ \frac{1}{2d^2}[\frac{ds}{r^2}+\arctan\frac{s}{d}], & k=0,\end{cases}\\
L^k &= \int \frac{d^3s^k}{r^6}\, ds = \begin{cases} \frac{d^2}{4}[-\frac{ds^{k-1}}{r^4}+(k-1)K^{k-2}], & k\ge 2,\\ \frac{d^3}{4r^4}, & k=1,\\ \frac{1}{4}[\frac{ds}{r^4}+3K^0], & k=0,\end{cases}\\
M^k &= \int \frac{d^5s^k}{r^8}\, ds = \begin{cases} \frac{d^2}{r^6}[-\frac{d^3s^{k-1}}{r^6}+(k-1)L^{k-2}], & k\ge 2,\\ \frac{d^5}{6r^6}, & k=1,\\ \frac{1}{6}[\frac{d^3s}{r^6}+5L^0], & k=0.\end{cases}
\end{aligned}
$$

All these integrals are indefinite integrals. To apply these formulas to an integral over a given element $\Gamma_i \hat{=} [s_a\,,\,s_b]$ you, naturally, have to substitute the integration limits as in

$$
\int_{s_a}^{s_b} s^2 \ln r\, ds = \frac{1}{3}[s^3\ln r - J^4]_{s_a}^{s_b} \qquad \text{etc.}
$$

With these formulas an analytical integration of the regular boundary for any given order n of polynomial shape functions can be carried out. This facilitates the implementation of an adaptive boundary element procedure for the plate equation.

These analytical integration formulas for the regular boundary integrals has been implemented in the programm BE-PLATTE. A comparison with results obtained with Gaussian quadrature shows that there is no measurable gain of accuracy in the boundary values of the plate bending solution. Different versions of the code that work with analytical integration and numerical quadrature render nearly identical results. But the gain in accuracy is remarkable when we turn to the calculations of the bending moments at points close to the boundary. In Fig. 3 you see the distribution of the principal moments for a slightly skew-angled plate. The stress-lines run parallel to the x-axis and so at the upper and lower edge the stress-points come close to the boundary. Gaussian quadrature with ten points renders totally unreliable results. Fig. 4 shows the same results obtained with analytical integration. The gain in accuracy is remarkable.

About the calculation of these results we want to talk next.

Internal actions

The bending moments

$$\begin{aligned} M_{11} &= -K\,(w_{,11} + \nu w_{,22}\,), \qquad M_{12} = -K\,(1-\nu)w_{,12}\,, \\ M_{22} &= -K\,(w_{,22} + \nu w_{,11}\,), \end{aligned}$$

are linear combinations of the second derivatives

$$w_{,11}\,, \quad w_{,12}\,, \quad w_{,22}\,,$$

so that, instead of formulating three separate representation formulas for the three bending moments, it is more efficient to formulate only one representation formula for a second order directional derivative

$$\frac{\partial}{\partial m}\frac{\partial}{\partial n}w.$$

Let the angles α, β, γ denote the orientation of the directions $\nu, \mathbf{n}, \mathbf{m}$

$$\nu_1 = \cos\alpha, \qquad \nu_2 = \sin\alpha,$$

$$m_1 = \cos\beta, \qquad m_2 = \sin\beta,$$

$$n_1 = \cos\gamma, \qquad n_2 = \sin\gamma\,.$$

We express the second order derivative $\frac{\partial^2}{\partial m \partial n}$ by the 'element intrinsic operators' $\frac{\partial}{\partial \nu}$ and $\frac{\partial}{\partial \tau}$ as follows

$$\begin{aligned} \frac{\partial^2}{\partial m \partial n} =& \cos(\beta-\alpha)\cos(\gamma-\alpha)\frac{\partial^2}{\partial \nu^2} \\ &+ \sin(\beta+\gamma-2\alpha)\frac{\partial^2}{\partial \nu \partial \tau} + \sin(\beta-\alpha)\sin(\gamma-\alpha)\frac{\partial^2}{\partial \tau^2}. \end{aligned}$$

The application of this operator to the kernel functions becomes particularly simple if we note that

$$r_{\nu\nu} = \frac{1}{r}r_{,\tau}^2, \qquad r_{\nu\tau} = -\frac{1}{r}r_{,\nu}r_{,\tau}, \qquad r_{\tau\tau} = \frac{1}{r}r_{,\nu}^2.$$

Now to actually calculate the derivatives

$$\frac{\partial^2}{\partial\nu^2}, \qquad \frac{\partial^2}{\partial\nu\partial\tau}, \qquad \frac{\partial^2}{\partial\tau^2}$$

we use the kernels

$$(g_0(\mathbf{y},\mathbf{x}))_{,n\,,m} \quad (\frac{\partial}{\partial\nu}g_0(\mathbf{y},\mathbf{x}))_{,n\,,m} \quad (M_\nu(g_0(\mathbf{y},\mathbf{x})))_{,n\,,m} \quad (V_\nu(g_0(\mathbf{y},\mathbf{x})))_{,n\,,m}$$

given in [1]. By making the substitutions

$$\mathbf{n}=\nu \quad \mathbf{m}=\nu \quad r_n=-r_\nu \quad r_t=-r_\tau \quad r_m=-r_\nu \quad r_p=-r_\tau$$

$$\mathbf{n}=\nu \quad \mathbf{m}=\tau \quad r_n=-r_\nu \quad r_t=-r_\tau \quad r_m=-r_\tau \quad r_p=r_\nu$$

$$\mathbf{n}=\tau \quad \mathbf{m}=\tau \quad r_n=-r_\nu \quad r_t=r_\nu \quad r_m=-r_\tau \quad r_p=r_\nu$$

(note that the two vectors τ (= direction 1) and ν (= direction 2) must form a right-handed system) we obtain the following results

$$\begin{aligned}
\frac{\partial^2}{\partial\nu^2}g_0(\mathbf{y},\mathbf{x}) =& \frac{1}{8\pi K}(1+2\ln r+2r_\nu^2)\\
\frac{\partial^2}{\partial\tau^2}g_0(\mathbf{y},\mathbf{x}) =& \frac{1}{8\pi K}(1+2\ln r+2r_\tau^2)\\
\frac{\partial^2}{\partial\nu\partial\tau}g_0(\mathbf{y},\mathbf{x}) =& \frac{1}{4\pi K}r_\nu r_\tau
\end{aligned}$$

$$\begin{aligned}
\frac{\partial^2}{\partial\nu^2}\frac{\partial}{\partial\nu}g_0(\mathbf{y},\mathbf{x}) =& \frac{1}{4\pi Kr}(1+2r_\nu^2)r_\nu\\
\frac{\partial^2}{\partial\tau^2}\frac{\partial}{\partial\nu}g_0(\mathbf{y},\mathbf{x}) =& \frac{1}{4\pi Kr}(r_\nu^2-r_\tau^2)r_\nu\\
\frac{\partial^2}{\partial\nu\partial\tau}\frac{\partial}{\partial\nu}g_0(\mathbf{y},\mathbf{x}) =& -\frac{1}{4\pi Kr}(r_\nu^2-r_\tau^2)r_\tau,
\end{aligned}$$

$$\begin{aligned}
\frac{\partial^2}{\partial\nu^2}M_\nu(g_0(\mathbf{y},\mathbf{x})) =& \frac{1}{4\pi r^2}([(1+\nu)+2(1-\nu)r_\tau^2](r_\nu^2-r_\tau^2)\\
&+4(1-\nu)r_\nu^2r_\tau^2)\\
\frac{\partial^2}{\partial\nu\partial\tau}M_\nu(g_0(\mathbf{y},\mathbf{x})) =& \frac{1}{4\pi r^2}[2(1+\nu)+4(1-\nu)(r_\tau^2-r_\nu^2)]r_\nu r_\tau\\
\frac{\partial^2}{\partial\tau^2}M_\nu(g_0(\mathbf{y},\mathbf{x})) =& \frac{1}{4\pi r^2}([1+\nu)-2(1-\nu)r_\tau^2](r_\tau^2-r_\nu^2)\\
&+4(1-\nu)r_\nu^2r_\tau^2),
\end{aligned}$$

$$\begin{aligned}
\frac{\partial^2}{\partial\nu^2}V_\nu(g_0(\mathbf{y},\mathbf{x})) =& \frac{1}{4\pi r^3}\{2r_\nu\{[3-\nu-2(1-\nu)r_\tau^2](r_\tau^2-r_\nu^2)\\
&+4(1-\nu)r_\nu^2r_\tau^2\}-r_\tau\{12(1-\nu)r_\nu r_\tau(r_\tau^2-r_\nu^2)\\
&-4[3-\nu-2(1-\nu)r_\tau^2]r_\nu r_\tau\}\},
\end{aligned}$$

$$
\begin{aligned}
\frac{\partial^2}{\partial\tau^2}V_\nu(g_0(\mathbf{y},\mathbf{x})) = &\frac{1}{4\pi r^3}\{2r_\tau\{-2[3-\nu-2(1-\nu)r_\tau^2]2r_\nu r_\tau \\
&-4(1-\nu)r_\nu^3 r_\tau\} + r_\nu\{-8(1-\nu)r_\nu^2 r_\tau^2 \\
&-2[3-\nu-2(1-\nu)r_\tau^2](r_\tau^2 - r_\nu^2) \\
&+4(1-\nu)r_\nu^2(r_\nu^2 - 3r_\tau^2)\}\} \\
\frac{\partial^2}{\partial\nu\partial\tau}V_\nu(g_0(\mathbf{y},\mathbf{x})) = &\frac{1}{4\pi r^3}\{2r_\tau\{[3-\nu-2(1-\nu)r_\tau^2](r_\tau^2 - r_\nu^2) \\
&+4(1-\nu)r_\nu^2 r_\tau^2\} + r_\nu\{4(1-\nu)r_\nu r_\tau(r_\tau^2 - r_\nu^2) \\
&-2[3-\nu-2(1-\nu)r_\tau^2]2r_\nu r_\tau \\
&+4(1-\nu)(2r_\nu r_\tau^3 - r_\nu^3 r_\tau - r_\nu^3 r_\tau)\}\}.
\end{aligned}
$$

Then we replace r_ν by $\frac{d}{r}$ and r_τ by $\frac{s}{r}$. The integrals of the transformed kernels times their boundary layers can also be expressed as linear combinations of the five integrals I^k, J^k, K^k, L^k and M^k.

The calculation of the shear forces is done in a similiar way. The differential operator

$$\frac{\partial^3}{\partial l\partial m\partial n}$$

is expressed by the 'element intrinsic operators' $\frac{\partial}{\partial\nu}$ and $\frac{\partial}{\partial\tau}$ as follows

$$
\begin{aligned}
\frac{\partial}{\partial l}\frac{\partial}{\partial m}\frac{\partial}{\partial n} = &\cos(\delta-\alpha)\cos(\beta-\alpha)\cos(\gamma-\alpha)\frac{\partial^3}{\partial\nu^3} \\
&+[\cos(\delta-\alpha)\cos(\beta-\alpha)\sin(\gamma-\alpha) \\
&+\cos(\delta-\alpha)\sin(\beta-\alpha)\cos(\gamma-\alpha) \\
&+\sin(\delta-\alpha)\cos(\beta-\alpha)\cos(\gamma-\alpha)]\frac{\partial^3}{\partial\nu^2\partial\tau} \\
&+[\cos(\delta-\alpha)\sin(\beta-\alpha)\sin(\gamma-\alpha) \\
&+\sin(\delta-\alpha)\cos(\beta-\alpha)\sin(\gamma-\alpha) \\
&+\sin(\delta-\alpha)\sin(\beta-\alpha)\cos(\gamma-\alpha)]\frac{\partial^3}{\partial\nu\partial\tau^2} \\
&+\sin(\delta-\alpha)\sin(\beta-\alpha)\sin(\gamma-\alpha)\frac{\partial^3}{\partial\tau^3},
\end{aligned}
$$

where the angle δ denotes the orientation of the direction $\mathbf{l}$. The following steps are still to be done.

References

[1] Hartmann, F.:
Introduction to Boundary Elements
Springer 1989.

[2] Hartmann, F.:
The Mathematical Foundation of Structural Mechanics
Springer 1985.

[3] Hartmann, F.:
ELASTOSTATICS
in: Brebbia, C.A.: Progress in BOUNDARY ELEMENT METHODS,
Pentech Press 1981.

[4] Blum, H.:
Numerical Treatment of Corner and Crack Singularities
Saarbrücken, 1987.

[5] Stern, M.:
A general boundary integral formulation
for the numerical solution of plate bending problems
Int. J. Solids Structures **15** (1979) pp. 769 - 782.

[6] Melzer, H.; Rannacher, R.:
Spannungskonzentrationen in Eckpunkten der Kirchhoffschen Platte
Bauingenieur 55 (1980) 181-184.

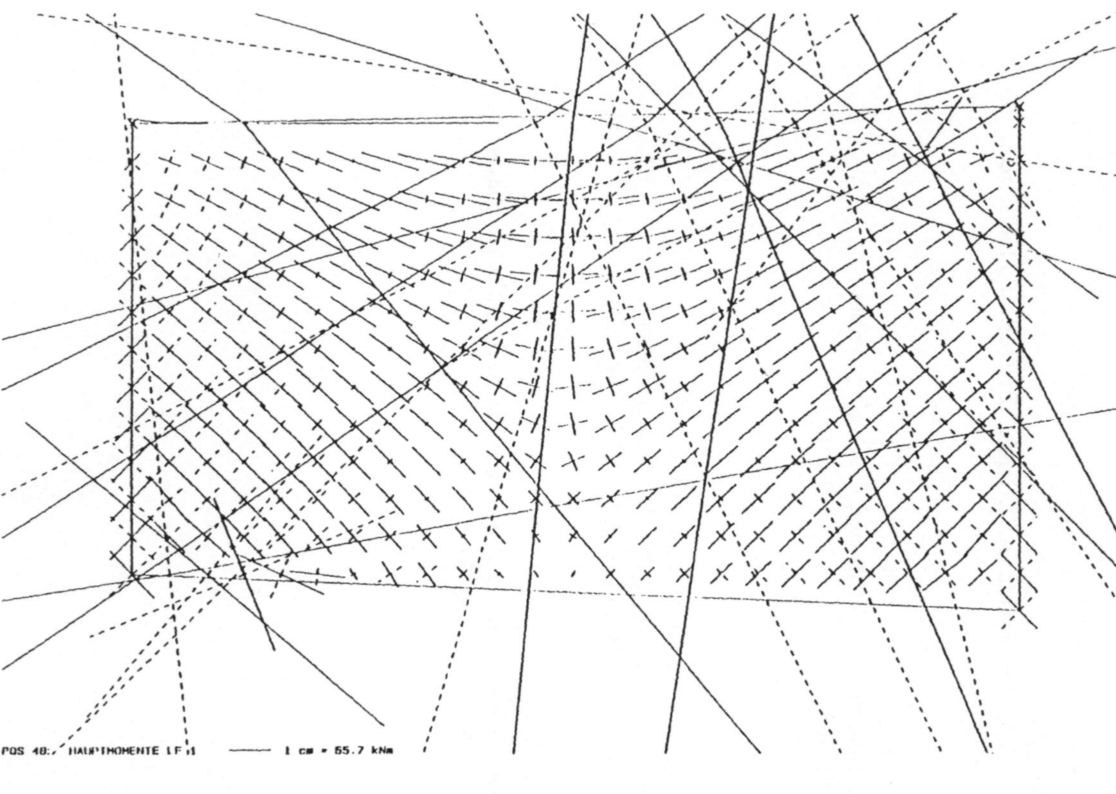

Fig.3 distribution of principal moments
(calculated by Gaussian quadrature)

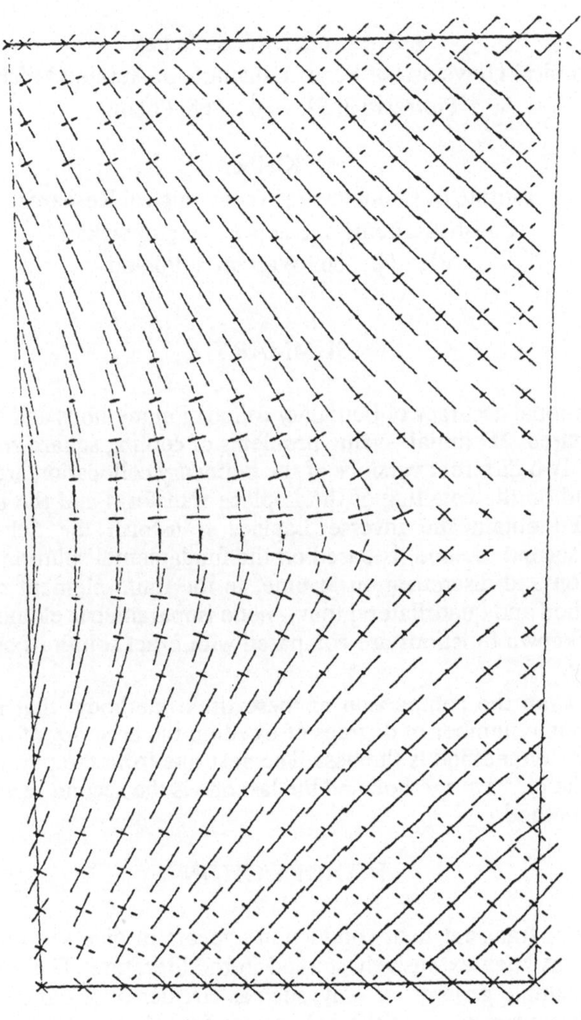

Fig.4 distribution of principal moments
(calculated by analytical integration)

TRANSIENT HEAT CONDUCTION BY BOUNDARY COLLOCATION METHODS AND FEM – A COMPARISON STUDY

J. A. Kołodziej, J. Stefaniak
Technical University of Poznan, Institute of Applied Mechanics
Piotrowo 3, PL–60–965, Poland

M. Kleiber
Institute of Fundamental Technological Research
Polish Academy of Science, Swiętokrzyka 21
PL–00–069 Warszawa, Poland

SUMMARY

The computational accuracy of boundary collocation methods and the finite element method is compared. 2D initial–value problems of cooling square rectangular regions are considered. Two different versions of the boundary collocation are considered. The first one is based on the coupling of the Laplace transform and the boundary collocation method. To obtain the inverse Laplace transform the Schapery method is employed. The second method is based on the fundamental solution of governing differential equation and discretization in time. In the finite element method the backward Euler method and quadrilateral four–node isoparametric elements are used. The solutions for unknown functions are compared with exact solutions as functions of the mesh refinement.

It follows from the comparison of these three methods that in considered examples, for the same number of degrees of freedom the accuracy of the first version of boundary collocation method is the best. The next one from the point of view of its accuracy is the finite element method and the last one is the second version of the boundary collocation method.

INTRODUCTION

The problem of transient heat conduction appears in diverse areas of science and technology and has been extensively studied in the literature. The analytical solutions are possible for simple geometries only, such as circular or rectangular regions. Thus, numerical methods provide often the only means for solving transient heat conduction problems of engineering significance. The techniques most widely used are the finite difference method(FDM) and the finite element method (FEM) . Treatment of complicated boundaries in the FDM has always been difficult and inaccurate; therefore researches have in most cases resorted to the FEM. Another method which is appropriate for complicated boundaries and which has gained popularity in the last decade is the boundary integral equations method and, particularly, its version called the boundary element method.

Using boundary methods for solving problems of transient heat conduction runs into some difficulties. A general feature of every boundary method is the strict fulfillment of the governing differential equations within the region considered while the boundary conditions are satisfied in an approximate way only. The boundary methods have basically been employed to solve boundary–value problems rather than initial–boundary–value problems. Using a boundary method for problems in the latter group requires a specific time discretization or time transformation which introduce additional errors into the solution.

It has been recently demonstrated that the so called boundary collocation method (BCM), which belongs to the general class of the boundary methods, may have some advantages over FEM (measured in terms of the solution accuracy) when used to solve some test examples of harmonic boundary–value problems, [1–2]. In this paper we address the similar problem of accuracy assessment of the boundary collocation method vs FEM by solving a number of test examples describing nonstationary heat conduction (i.e. initial–boundary–value problems).

Three 2–D initial–value problems are considered for which exact solutions are available. The main question to be posed below is which of the two methods yields more accurate results given the same level of discretization measured by the number of degrees of freedom assumed, and for which one the computation time is smaller.

TEST PROBLEMS AND THE ANALYTICAL SOLUTIONS

The problems chosen for this study in dimensionless forms are as follows
Problems I, II and III

$$\frac{\partial^2 \theta}{\partial X^2} + \frac{\partial^2 \theta}{\partial Y^2} = \frac{\partial \theta}{\partial T} \quad \text{in } 0 < X < 1,\ 0 < Y < E \tag{1}$$

with the initial and boundary conditions

$$\theta = 0 \quad \text{at} \quad T = 0 \quad \text{for} \quad 0 \le X \le 1,\ \ 0 \le Y \le E \tag{2}$$

$$\theta = -1 \quad \text{at} \quad T > 0 \quad \text{for} \quad X = 1, \quad 0 < Y < E \tag{3}$$

$$\theta = -1 \quad \text{at} \quad T > 0 \quad \text{for} \quad Y = \text{E}, \quad 0 < X < 1 \tag{4}$$

$$\frac{\partial \theta}{\partial Y} = 0 \quad \text{at} \quad T > 0 \quad \text{for} \quad Y = 0, \quad 0 < X < 1 \tag{5}$$

$$\frac{\partial \theta}{\partial X} = 0 \quad \text{at} \quad T > 0 \quad \text{for} \quad X + 0, \quad 0 < Y < E \tag{6}$$

where θ is the dimensionless temperature, T is the dimensionless time (Fourier's number), X and Y are the dimensionless coordinates.

The values of $E = 1.0$, $E = 0.5$ and $E = 0.25$ characterize problems I. II and III, respectively.

The exact solution of this problem has the form

$$\theta = -1 + \frac{16}{\pi^2} \sum_{m=1}^{\infty} \frac{(-1)^{m+n+2} \cos[(2m-1)\pi X/2] \cos[(2n-1)\pi EY/2}{(2m-1)(2n-1)} *$$

$$* \exp\left\{-[(2m-1)^2 + (2n-1)^2 E^2]\,\pi^2 T/4\right\}. \tag{7}$$

THE BOUNDARY COLLOCATION METHOD FOR INITIAL–BOUNDARY VALUE PROBLEMS

Two different versions of boundary collocation are considered. The first one is based on the Laplace transform of governing differential equations and of boundary conditions, [3,4], and will be referred to below as BCMLT. The second method to be considered is based on the fundamental solution of the governing differential equations and the discretization of time variable, [5]. Here, this method is called the boundary collocation method with the fundamental solutions (BCMFS).

The boundary collocation with Laplace transformation method

The main idea of the BCMLT is as follows. Applying the Laplace transformation with respect to time to eq. (1) and using the zero initial conditions in all the problems we have

$$\left(\frac{\partial^2}{\partial X^2} + \frac{\partial^2}{\partial Y^2}\right) \tilde{\theta}(X,Y,p) = p\,\tilde{\theta}(X,Y,p) \quad \text{in } 0 < X < 1,\ 0 < Y < E \tag{8}$$

where

$$\tilde{\theta}(X,Y,p) = \int_0^\infty e^{-pT}\,\theta(X,Y,T)\,dT.$$

Applying the Laplace transform to eqs. (3–6) we have

$$\tilde{\theta} = -1/p \quad \text{for} \quad X = 1, \quad 0 < Y < E \tag{9}$$

$$\tilde{\theta} = -1/p \quad \text{for} \quad Y = E, \quad 0 < X < 1 \tag{10}$$

$$\frac{\partial\tilde{\theta}}{\partial Y} = 0 \quad \text{for} \quad Y = 0, \quad 0 < Y < E \tag{11}$$

$$\frac{\partial\tilde{\theta}}{\partial X} = 0 \quad \text{for} \quad X = 0, \quad 0 < Y < E. \tag{12}$$

In this way the initial–boundary value problems (1–6) have been transformed to boundary value problems (8–12). In the conventional formulation of BEM to solution of boundary–value problems, the singularities of fundamental solutions are located on the boundary of the problem. This practice presents some drawbacks. For example, special attention is required to evaluate singular integrals. Kupradze [6] proposed locating the boundary nodes on an auxiliary boundary which encloses the boundary of the real problem.

According to the Kuprodze's idea, the solution of eq. (8) may be sought in the form

$$\tilde{\theta} = \frac{1}{2\pi}\oint_\Gamma \sigma(p,\xi,\eta)\, K_0\left\{\left[p\,((X-\xi)^2 + (Y-\eta)^2)\right)\right]^{1/2}\right\} d\Gamma \tag{13}$$

where Γ is the auxiliary boundary, σ is the unknown density, ξ,η are the coordinates of points on auxiliary boundary, K_0 is the modified Besselfunction. If the integral (13) is approximated by finite sum, in test problems considered (accounting for symmetry) the assumed solution has the form

$$\tilde{\theta} = \frac{1}{2\pi} \sum_{j=1}^{N} a_j(p)\, K(p, X, Y, \xi_j, \eta_j) \tag{14}$$

where

$$K(p,X,Y,\xi_j,\eta_j) = K_0\{[p\,((X-\xi_j)^2+(Y-\eta_j)^2))^{1/2}\} +$$

$$+ K_0\{[p\,((X+\xi_j)^2+(Y-\eta_j)^2))]^{1/2}\} + K_0\{[p\,((X-\xi_j)^2+(Y+\eta_j)^2))]^{1/2}\} +$$

$$+ K_0\{[p\,((X+\xi_j)^2+(Y+\eta_j)^2))]^{1/2}\}\,.$$

In the last formula (ξ_j , η_j) are coordinates source points placed outside the considered region , see Fig. 1, while a_j are unknown parameters to be found from the boundary conditions. In this way the symmetry conditions (11–12) are fulfill exactly.

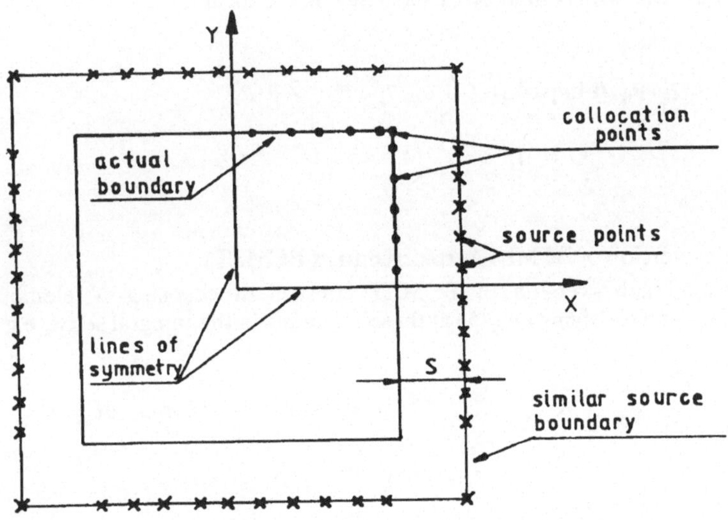

Fig. 1. Set–up for BCMLT and BCMFS.

The problem now is to determine the parameters a_j and to invert the Laplace transform.

The problem of inverting a general transform function known only numerically is treated in the survey article [7]. According to its authors it is clear that no existing method of an approximate inversion can be expected to provide acceptable results for completely arbitrary transform functions. However, under certain reasonable restrictions on the character of the transform functions, or perhaps on the time function sought, excellent results are obtainable by the schemes given in literature .Considering the heat conduction problems, we shall adopt the inversion method of Schapery [8] proposed for viscoelastic stress analysis.

The method consists in selecting a finite sequence of the real values of transform parameters p_1 , p_2 ,... p_M . The way of choosing these parameters will be described below. By consequently substituting (14) taken at values $p_1, p_2, \ldots p_M$ into the boundary

below. By consequently substituting (14) taken at values $p_1, p_2, ... p_M$ into the boundary condition (9–10) we arrive at M relations of the form

$$\sum_{j=1}^{N} a_{jk} \; K(p_k, X^0, Y^0, \xi_j, \eta_j) = -2\pi/p_k \tag{15}$$

$$k = 1, 2, \dots, M$$

where X^0, Y^0 are coordinates of points on boundary $X = 1$, $0 < Y < E$ and $Y = E$, $0 < X < 1$. Since the values of constants a_j depend on the selected parameter values p_k we introduce the notation a_{jk} for a_j.

Several possibilities exist for satisfying eq. (15). The simplest one consists in choosing N points with coordinates (X_i^0, Y_i^0) along the boundary iconsidered and anforcing eq. (15) at these points. In this way we may obtain M sets of N linear equations with N unknowns, each having the form

$$\sum_{j=1}^{N} a_{jk} \; K(p_k, X_i^0, Y_i^0, \xi_j, \eta_j) = -2\pi/p_k \tag{16}$$

$$k = 1, 2, \dots, M$$
$$i = 1, 2, \dots, N .$$

This sub–version of BCMLT is referred to as BCMLT1.

The second sub–version of BCMLT consists in choosing N elements E_i on the boundary and satisfying eq.(15) at these elements in the integral sense, e.g.:

$$\sum_{j=1}^{N} a_{jk} \int_E K(p_k, X^0, Y^0, \xi_j, \eta_j)\, dE = \int_E -2\pi/p_k \; dE_i \tag{17}$$

$$k = 1, 2, \dots, M$$
$$i = 1, 2, \dots, N .$$

This sub–version of BCMLT is referred to as BCMLT2.

In the third sub–version, referred to as BCMLT3, the integral conditions are fulfilled in the least square sense.

Gauss elimination was employed in this study to solve eq. (16) or (17) or (17) in the least square sense for the unknowns a_{jk}.

Having computed a_{jk} the solution in the considered domain is assumed in the form

$$\theta(X, Y, T) = -1 + \sum_{l=1}^{M} c_l (X, Y) \exp(-b_l T) \tag{18}$$

where $c_l(X,Y)$ are unknown functions while b_l unknown scalars.

Laplace transform applied to the last formulae gives

$$\widetilde{\theta}(X,Y,p) = -\frac{1}{p} + \sum_{l=1}^{M} \frac{c_l(X,Y)}{p+b_l}. \tag{19}$$

Comparing the R.H.S.'s of eqs. (14) and (19) leads to

$$-\frac{1}{p} + \sum_{l=1}^{M} \frac{c_l(X,Y)}{p+b_l} = \frac{1}{2\pi} \sum_{j=1}^{N} a_j(p)\, K(p, X, Y, \xi_j, \eta_j). \tag{20}$$

We assume that b_l are equal to p_l and the coordinates X_k, Y_k are given specific values, then the last relationships reads as

$$\mathbf{Ac} = \mathbf{B} \tag{21}$$

which is a set of linear equations with the unknowns c_l whereas

$$A_{ij} = \frac{1}{p_i + p_j} \qquad i, j = 1, 2, \ldots, M \tag{22}$$

$$B_i^{(k)} = \frac{1}{p_i} + \frac{1}{2\pi} \sum_{j=1}^{N} a_{ji}\, K_0 \left\{[p_0((X_k - \xi_j)^2 + (Y_k - \eta_j)^2))]^{1/2}\right\} \tag{23}$$

$$c_j^{(k)} = c_j(X_k, Y_k).$$

The fundamental boundary collocation method

This method is presented in detail in [5]. Its main idea can be summarized as follow. Let us assume, that concentrated heat sources of kintensities W_i^k act at points (ξ_i, η_i) outside the considered region (see Fig. 1) and change stepwise in time. The temperature caused by this sources and consequently the solution of the initial – boundary value problems (1 – 6) is given by the formula [5]

$$\theta(X,Y,T) \sum_{i=1}^{N} \sum_{k=0}^{M} W_i^k\, G(X-\xi_i, Y-\eta_i, T-T_k)\, \eta(T-T_k) \tag{24}$$

where W_i^k are arbitrary constants to be found from the boundary conditions, and

$$G = \frac{1}{4\pi} \left\{ \mathrm{Ei}\left(-\frac{[(X-\eta_i)^2 + (Y-\eta_i)^2]}{\sqrt{T-T_k}}\right) + \mathrm{Ei}\left(-\frac{[(X+\xi_i)^2] + (Y-\eta_i)^2]}{\sqrt{T-T_k}}\right) + \right.$$

$$\left. + \mathrm{Ei}\left(-\frac{[(X-\xi_i)^2 + (Y+\eta_i)^2]}{\sqrt{T-T_k}}\right) + \mathrm{Ei}\left(-\frac{[(X+\xi_i)^2 + (Y+\eta_i)^2]}{\sqrt{T-T_k}}\right)\right\}$$

where

$$\mathrm{Ei}(-z) = -\int_z^\infty \frac{1}{u} \exp(-u)\, du.$$

The temperature given by the formula (24) satisfies eq. (1), the initial condition (2) and the symmetry conditions (5–6). Inserting the solutions (24) into the boundary conditions (3–4) at N points (X_j^0, Y_j^0) and at moments T_k the following M sets of N linear algebraic equations for $M*N$ unknown values W_i^k are obtained:

$$\sum_{i=1}^{N} \sum_{k=0}^{l-1} W_i^k \; G(X_j^0 - \xi_i, Y_j^0 - \eta_i, T_l - T_k) \; \eta(T_l - T_k) = -1 \tag{25}$$

$$l = 1, 2,..,M$$

$$j = 1, 2,..., N .$$

THE FINITE ELEMENT METHOD

The finite element technique employed in this study has been described in [9]. The quadrilateral four–node or eight–node isoparametric elements are used in the computer program. The matrix equation describing the transient heat conduction has the form

$$C \,\theta + K \,\theta = Q \tag{26}$$

in which C and K are the global heat capacity and conductivity matrices, Q is thermal 'load' vector and q is the nodal temperature vector. The unconditionally stable Euler method is used to solve eq. (26).

For the test problems considered discretization meshes are shown in Fig. 2.

ERROR CRITERIA

Two different error criteria have been employed. The first one is based on 'global' error measure for the dimensionless temperature and is given by

$$ERG = \frac{1}{NT*NP} \sum_{i=1}^{NT} \sum_{j=1}^{NP} \mid \theta_e(X_j, Y_j, T_i) - \theta_a(X_j, Y_j, T_i, N) \mid . \tag{27}$$

The subscripts e and a above refer to the exact and approximate solutions obtained by means of either BCM or FEM, respectively. Points (X_j, Yj) at which BCM errors are evaluated are uniformly distributed over regions . FEM errors are evaluated at nodes. In BCM the number NP is equal to 15 or 25 but in the FEM the number NP stands for the number of nodes and depend on the discretization meshes. Equal time steps are selected with the temperature T_1 at (X_1, Y_1) and the first time instant made not to exceed 10 % of the boundary temperature and the temperature T_{NT}at (X_1, Y_1) made exceed 90 % of this value. The time instants are thus different for each problem analyzed but they 'encompass' the whole range of the dimensionless temperature. Thus the global error measure is the average deviation of the approximate solution from the exact solution in space and time.

The second error criterion has a local character and is defined by

$$ERL = \max \mid \theta_e(X_j, Y_j, T_i) - \theta_a(X_j, Y_j, T_i, N) \mid . \tag{28}$$

To simplify the FEM computations, the maximum is taken over the time steps and nodes in the finite element mesh.

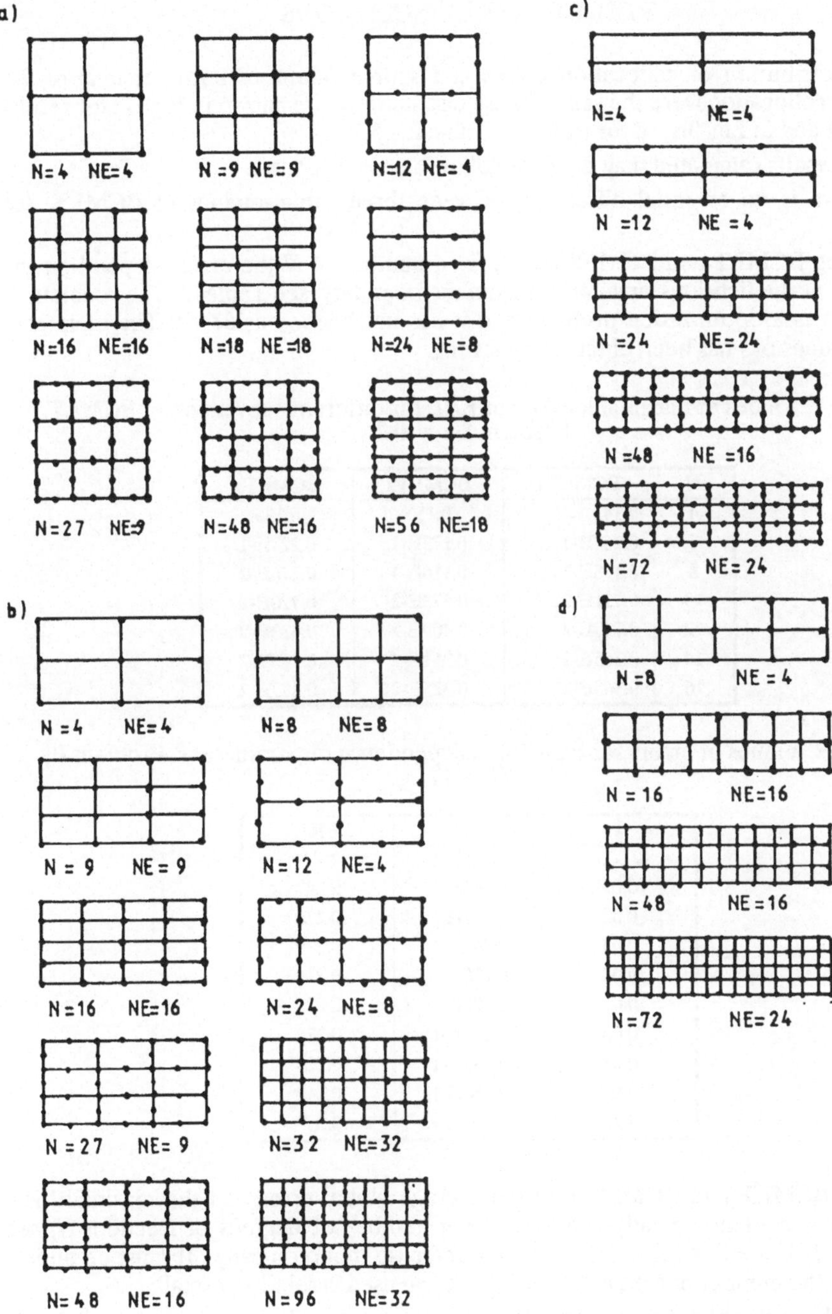

Fig. 2. Discretization meshes for FEM, N = number of degrees of freedom, NE = number of elements. a) Problem I, b) Problem II, c) Problem III.

RESULTS AND CONCLUSIONS

The distribution of collocation points and source points in both versions of the boundary collocation were the same. These distributions are shown in Fig. 3a for results of Table 1 and in Fig. 3b – d for results of Table 2 – 6.

From results calculated may draw the following conclusions:

1. There is no essential difference between three sub – versions of BCMLT, see Table 1.

2. Using BCMLT and BCMFS with a fixed number of N the errors depend on the distance s of the fictitious singularities from the boundary, see Table 2. The problem of choosing s is a optimization problem and it has not been considered. The most convenient values of s has been obtained by testing.

Table 1. Values of maximal local error *ERL* for different subversion of *BCMLT*. Problem I. $s = 0.2$.

N	BCMLT1	BCMLT2	BCMLT3
4	0.145	0.153	0.135
6	0.267E-1	0.173E-1	0.227E-1
8	0.178E-1	0.116E-1	0.276E-2
10	0.731E-2	0.573E-2	0.720E-2
12	0.727E-2	0.409E-2	0.433E-2
14	0.365E-2	0.311E-2	0.185E-2
16	0.365E-2	0.325E-2	0.377E-2

Table 2. Values of errors for BCMFS in dependence of parameter s. Problem II, $N = 17$

s	ERG	ERL
0.005	0.656E-1	0.272
0.01	0.483E-1	0.247
0.02	0.331E-1	0.226
0.03	0.261E-1	0.219
0.04	0.225E-1	0.219
0.05	0.205E-1	0.221
0.06	0.194E-1	0.225
0.07	0.189E-1	0.229
0.09	0.186E-1	0.240
0.12	0.189E-1	0.257

3. For BCMLT and BCMFS both the average global error and the maximal local error decrease monotonically with increasing number of degrees of freedom N, see Tables 3 – 5. For FEM these errors do not decrease monotonically . It must be underlined, that the number of elements which has been used is relatively small.

4. In all the methods the maximal local errors is at the first time step, see Table 6. The reason is probably that we have a jump of temperature at the boundary at $t = 0$. The BCMFS method is particularly sensitive to the time discontinuity in the boundary condition.

Table 3. Comparison of errors for FEM and BCM. Problem I.Time step for FEM $DT = 0.05$, number of time step $NTS = 20$

N	FEM		BCMLT1		BCMFS	
	ERG	ERL	ERG	ERL	ERG	ERL
			s=0.2		s=0.04	
4	0.186E-1	0.716E-1				
7			0.208E-2	0.941E--2	0.668E-1	0.385
9	0.175E-1	0.112E+0	0.743E-3	0.759E-2	0.507E-1	0.215
11			0.244E-3	0.536E-3	0.335E-1	0.247
12	0.144E-1	0.977E-1				
13			0.139E-3	0.222E-2	0.269E-1	0.234
15			0.990E-4	0.121E-2	0.236E-1	0.225
16	0.176E-1	0.165E+0				
17			0.927E-4	0.189E-3	0.225E-1	0.219
18	0.175E-1	0.163E-1				
19			0.865E-4	0.121E-2	0.198E-1	0.215
23			0.852E-4	0.121E-2	0.182E-1	0.211
24	0.158E-1	0.170E+0				
25			0.857E-4	0.121E-2	0.180E-1	0.210
27	0.162E-1	0.166E+0	0.854E-4	0.121E-2	0.174E-1	0.209
31					0.170E-1	0.209
48	0.172E-1	0.195E+0				
56	0.172E-1	0.192E+0				
64	0.184E-1	0.169+0				

5. In all the considered cases ($E = 1; 0.5; 0.25$) the best results has been obtained by the BCMLT, see Tables 3–5. This concerns both maximal local error and average global error. The results obtained by BCMFS are acceptable, but worse then these obtained by FEM.

6. In the light of the above examples the numerically most effective (in terms of the computational time) approach is the FEM followed by the BCMFS and, finally, by the BCMLT.

Table 4. Comparison of errors for FEM and BCM. Problem II. Time step for FEM $DT = 0.02$, number of time step $NTS = 20$

N	FEM		BCMLT1		BCMFS	
	ERG	ERL	ERG	ERL	ERG	ERL
4	0.178E-1	0.865E-1				
7			0.160E-2	0.316E-1	0.517E-1	0.271
8	0.168E-1	0.698E-1				
10			0.427E-3	0.645E-2	0.485E-1	0.263
12	0.164E-1	0.125E+0				
13			0.275E-3	0.362E-2	0.236E-1	0.224
16	0.174E-1	0.144E+0	0.261E-3	0.362E-2	0.218E-1	0.223
19			0.255E-3	0.362E-2	0.189E-1	0.223
22			0.256E-3	0.362E-2	0.182E-1	0.222
27	0.171E-1	0.163E+0			0.173E-1	0.222
32	0.178E-1	0.174E+0				
48	0.154E-1	0.197E+0				

Table 5. Comparison of errors for FEM and BCM. Problem III. Time step for FEM $DT = 0.00625$, number of time step $NTS = 20$

N	FEM		BCMLT1		BCMFS	
	ERG	ERL	ERG	ERL	ERG	ERL
4	0.170E-1	0.570E-1				
8	0.167E-1	0.805E-1				
11			0.367E-2	0.111E+0	0.201E-1	0.300
12	0.286E-1	0.221E+0				
16	0.169E-1	0.104E+0	0.215E-2	0.640E-1	0.192E-1	0.245
21			0.154E-2	0.406E-1	0.182E-1	0.232
24	0.175E-1	0.141E+0				
26			0.121E-2	0.327E-1	0.177E-1	0.229
31			0.102E-2	0.275E-1	0.175E-1	0.227
36	0.177E-1	0.170E+0	0.894E-3	0.237E-1		
41			0.784E-3	0.209E-1		

Table 6. Values of maximal local error ERL in space as a function of time T. Problem I. $N = 9$.

T	BCMFS	BCMLT1	FEM
0.05	0.271	0.795E-2	0.112
0.10	0.170	0.759E-2	0.107
0.15	0.118	0.130E-2	0.735E-1
0.20	0.104	0.224E-2	0.512E-1
0.25	0.101	0148E-2	0.562E-1
0.30	0.929E-1	0.193E-2	0.637E-1
0.35	0.834E-1	0.249E-2	0.672E-1
0.40	0.739E-1	0.225E-2	0.657E-1
0.45	0.652E-1	0.163E-2	0.615E-1
0.50	0.574E-1	0.988E-3	0.561E-1
0.55	0.505E-1	0.436E-3	0.502E-1
0.60	0.446E-1	0.411E-3	0.444E-1
0.65	0.396E-1	0.504E-3	0.388E-1
0.70	0.353E-1	0.568E-3	0.336E-1
0.75	0.316E-1	0.600E-3	0.290E-1
0.80	0.285E-1	0.604E-3	0.248E-1
0.85	0.268E-1	0.583E-3	0.212E-1
0.90	0.261E-1	0.543E-3	0.179E-1
0.95	0.255E-1	0.491E-3	0.152E-1
1.00	0.249E-1	0.431E-3	0.128E-1

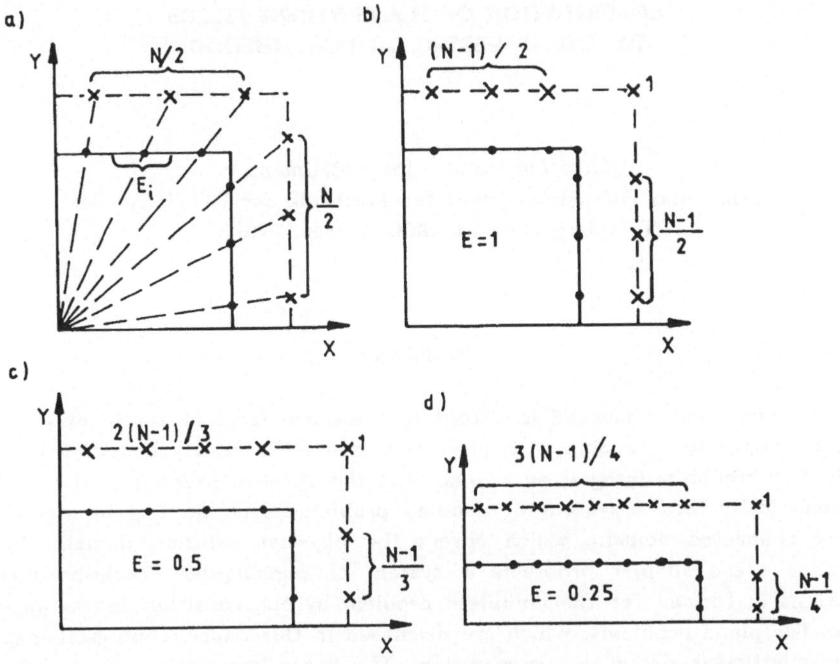

Fig. 3. The distribution of collocation and source points in both versions of BCM. *N* number degree of freedom

REFERENCES

[1] KOŁODZIEJ, J. A., KLEIBER, M., MUSIELAK, G.: "Comparison of the boundary collocation and finite element methods for some harmonic 2D problems", Mechanika Teoretyczna i Stosowana 26 (1988) pp. 647–666.

[2] KOŁODZIEJ, J. A., KLEIBER, M.: "Boundary collocation method vs FEM for some harmonic 2–D problems", Comput. Struct. 33 (1989) pp. 155–168.

[3] LO, C. F.: "Numerical solution the unsteady heat equation", AIAA Journal 7 (1969) pp. 973–975.

[4] TAKEUTI, Y., NODA, N.: "Transient thermoelastic problem in a polygonal cylinder with a circular hole", J. Appl. Mech. 40 (1973) pp. 935–940.

[5] STEFANIAK, J.: "Controlling the concentrated sources in some problems of heat conduction", J. Tech. Phys. 26 (1985) pp. 349–358.

[6] KUPRADZE, V.: "On the approximate solution of problems in mathematical physics", Rus. Math. Surveys, 22 (1967) pp. 59–107 (Uspehi Matematiceskih Nauk, 22 (1967) pp. 59–107).

[7] DAVIES, B., MARTIN, B.: "Numerical inversion of the Laplace transform: a survey and comparison of methods", J. Comput. Physics 33, (1979) pp. 1–32.

[8] SCHAPERY, R. A.: "Approximate methods of transform inversion for viscoelastic stress analysis", Proceedings of the Truth U. S. National Congress on Applied Mechanics, vol. 2 (1962) pp. 1075–1085.

[9] KLEIBER, M., SŁUŻALEC, A. Sz.: "Numerical analysis of heat flow in flash welding", Arch. Mech. 35 (1983) pp. 687–699.

COMPUTATION OF PLANE STRESS FIELDS BY THE COVERING DOMAIN METHOD

Xiao Lin and Josef Ballmann
Lehr- und Forschungsgebiet für Mechanik der RWTH Aachen
Templergraben 64, 5100 Aachen, FRG

SUMMARY

The covering domain method is a tool to treat problems of static elasticity in multiply connected domains with piece-wise smooth arbitrary boundaries by the method of Fredholm integral equations. First the physical problem is decomposed mathematically into a system of coupled problems each holding in a different simply connected domain, which covers the physical solution domain. Having upgraded these simpler problems a system of nonsingular Fredholm integral equations is formed for the complete problem by superposition in the physical plane. For plane problems, which are discussed in this paper, conformal mapping is most efficient doing this preparation. The paper deals with the forming of kernel functions and influence coefficients using rational fractional functions for conformal mapping. Beyond an introductory example in order to explain the method of covering domains, applications are dealt with for a plate with an elliptic hole and a rectangular plate with a lip crack.

INTRODUCTION

The physical problem is to solve an example of plane static elasticity for a multiply connected finite or infinite plate using the method of Fredholm integral equations. Following an idea of Mikhlin [1], applied and modified by several other authors, e.g. [2-4], the solution is formed by superposition of solutions each being valid in a different simply connected region covering the physical domain. To formulate these simpler problems a decomposition is made such that each of the decomposed problems can be prepared for solution by a Fredholm integral equation using e.g. Muskhelishvili's function [5], or, if known, the exact solution. In order to transform the different decomposed problems into a related standard form different conformal mappings are necessary for the different regions.

As an introductory example for the covering domain method we refer to a simple problem shown in Fig. 1(a), where an infinite plate with two collinear cracks, denoted by C_1 and C_2, is considered. The half lengths of the cracks are a and c, the central distance is b, and the crack boundaries are subjected to given normal loads $f_1(x)$ and $f_2(x)$, respectively, which are symmetrically distributed along the upper and lower parts of the cracks. The solution must take on these load distributions as boundary conditions.

The first step of the solution procedure is to select the covering domains. Because two boundaries exist in the problem, we select two infinite plates A_1 and A_2 as covering domains. In the first one, A_1, there is only one crack L_1, which has the same length as C_1 in the original plate A, see Fig. 1(b). In the same way, there is only one crack L_2 in A_2, which has the same length as C_2, see Fig. 1(c). Generally, we can suppose separately distributed self-equilibrated normal forces p_1 over L_1 and p_2 over L_2, and easily find the stress distribution in each covering domain for the assumed loading.

The second step is to construct the system of integral equations by superposition. The sum will be a possible solution in the original plate A. Demanding them to satisfy the prescribed boundary conditions on C_1 and C_2, the system of integral equations is formed.

For convenience, we choose the coordinate systems in A_1 and A_2 at the same position as that in A. Firstly, let us see the superposition of σ_y in C_1. The contribution from A_1 is the assumed loading force $p_1(x)$. The contribution from A_2 can be developed by Westergaard's function, which is well-known in fracture mechanics. Suppose the upper and lower surfaces of the crack L_2 are subjected to a self-equilibrated force distribution $p_2(t)$ at $z=t$. Then the Westergaard function for an infinetesimal load element $p_2(t)\,dt$ reads

$$Z_I^{(2)}(z) = \frac{-p_2(t)\,dt\sqrt{c^2-(t-b)^2}}{\pi(z-t)\sqrt{(z-b)^2-c^2}}\,, \qquad z \in A_2\,. \tag{1}$$

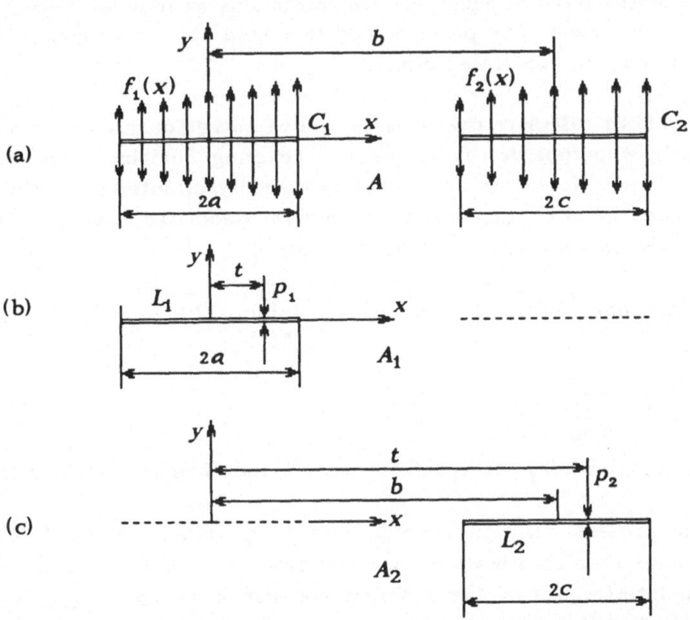

Fig. 1 Example with Two Colinear Cracks. (a) Original Plate; (b) Covering Domain A_1; (c) Covering Domain A_2.

For $z \in C_1$, namely for $-a<x<a$, $y=0$, the related stress contribution is

$$d\sigma_y^{(2)} = \frac{-p_2(t)dt\sqrt{c^2-(t-b)^2}}{\pi(t-x)\sqrt{(x-b)^2-c^2}} \,. \tag{2}$$

Integration over L_2 yields the total contribution from A_2. The superposition with the distributed load p_1 has to satisfy the boundary condition along C_1. So we have the equation:

$$p_1(x) - \int_{b-c}^{b+c} \frac{p_2(t)dt\sqrt{c^2-(t-b)^2}}{\pi(t-x)\sqrt{(x-b)^2-c^2}}\, dt = -f_1(x)\,, \quad (-a \le x \le a)\,. \tag{3a}$$

The procedure for C_2 is similar, involving a Westergaard funtion $Z_I^{(1)}$ and yielding the equation

$$p_2(x) - \int_{-a}^{a} \frac{p_1(t)\sqrt{a^2-t^2}}{\pi(x-t)\sqrt{x^2-a^2}}\, dt = -f_2(x)\,, \quad (-c \le x-b \le c)\,. \tag{3b}$$

The equations (3) represent a system of Fredholm integral equations of the second kind for the actual problem. Here we don't consider the effects of shear stress. Since the cracks are collinear and only subjected to normal forces, the shear stress in each covering domain vanishes along the x-axis.

The third step is to solve the integral equations with respect to $p_1(x)$ and $p_2(x)$. Generally, only a numerical solution will be possible. As can be seen from eq. (3) the kernels of the integral equations are continuous as long as $b-c>a$, i.e. if the two cracks don't meet. For problems of this kind the convergence in numerical computations can be controlled easily.

The last step is to calculate the stresses and the other quantities of interest; they will be got by superposition from the two covering domains. Here we solve for the stress intensity factor as an examplary quantity of interest at the point $z=a$. In the neighbourhood of this point, the stress associated with $Z_I^{(2)}$ is finite, so we find for the mode I stress intensity factor

$$K_I = \lim_{z\to a} \sqrt{2\pi(z-a)}\,\left(Z_I^{(1)} + Z_I^{(2)}\right) = \lim_{z\to a} \sqrt{2\pi(z-a)}\, Z_I^{(1)}$$

$$= -\frac{1}{\sqrt{\pi a}} \int_{-a}^{a} p_1(t)\sqrt{\frac{a+t}{a-t}}\, dt\,. \tag{4}$$

When $p_1(t)$ is solved, K_I follows from eq. (4) and then the work is finished.

We call the present solution procedure covering domain method (CDM). It is worthy to note that no iterations are necessary in the method to approximate the physical interference of the different boundaries as e.g. in [1]. The key idea of this method still belongs to the distributed source approach, namely, trying to represent the stress components by an integral of a distributed force $p_\alpha^k(t)$ over the boundary.

It should be mentioned that Chen [6] used an approach with superpositions of complex potentials, which turned out to be very efficient for problems with simple boundaries, like straight lines or circles. For more complicated boundaries, other approaches are still required.

For more general problems, there may exist n different boundaries in the physical problem. Then we can have n covering domains, numbered by $k = 1,..,n$. The solution for the k-th problem is to be evaluated to formulate its influence on all the other boundaries of the physical problem. The superposition of the contributions of all decomposed problems in the physical plane yields to a system of non-singular Fredholm integral equations. Unfortunately , for each formulation another conformal mapping plane can be involved and the inverse mappings can often not be achieved analytically. Then a numerical procedure is used after the working up of the integral equation of each decomposed problem in its mapping plane, in order to formulate its contribution to the complete solution in the physical plane by the so-called influence coefficients. Finally a system of linear equations results in the physical plane from the superposition which can be solved to give the solution of the complete physical problem.

It is evident that the most important part of CDM is the search for analytical solutions in the covering domains, with which the kernel functions can be formed. In [4] four important approaches were proposed on this topic, which are very efficient in dealing with technical problems of plates. Presently we will continue this work, with emphasis on the use of conformal mapping in the analysis.

We will focus the discussion on problems where the mapping function $z=\omega(\zeta)$ from the interior of the unit circle in the image plane into the physical solution domain is a rational fractional function. This of course is of a certain significance in engineering, because there always exists a conformal mapping function for any shape of a closed piece-wise smooth curve which is to be mapped onto the unit circle in the image plane. Expanding the mapping function into a Laurent series and taking finite terms for computation, sufficient accuracy can be attained.

FINITE PHYSICAL DOMAINS

Assume a finite domain in the physical plane (z-plane) where the solution is to be found, and A_k a finite covering domain in this plane. Let $\omega(\zeta)$ be a polynomial, which maps the interior of the unit circle in the ζ- plane (image plane) onto A_k in the z-plane.

$$z = \omega(\zeta) = c_1\zeta + c_2\zeta^2 + \cdots + c_n\zeta^n . \tag{5}$$

Assume a distributed traction $p = p_x + i\, p_y$ along the boundary ∂A_k of A_k, which is in equilibrium. According to [5], the stress components in the physical plane can be represented by the derivatives of two functions $\varphi(\zeta)$ and $\psi(\zeta)$ in the following way

$$\sigma_x + \sigma_y = 4\, Re\left[\frac{\varphi'(\zeta)}{\omega'(\zeta)}\right] \quad , \tag{6}$$

$$\sigma_y - \sigma_x + 2\tau_{xy} i = 2\left[\overline{\omega(\zeta)}\,\frac{\omega'(\zeta)\varphi''(\zeta) - \omega''(\zeta)\varphi'(\zeta)}{[\omega'(\zeta)]^3} + \frac{\psi'(\zeta)}{\omega'(\zeta)}\right] . \tag{7}$$

$\varphi(\zeta)$ and $\psi(\zeta)$ are analytic functions in the unit circle satisfying the following equations:

$$\varphi(\zeta) + \frac{1}{2\pi i}\oint_\gamma \frac{\omega(\sigma)}{\overline{\omega'(\sigma)}}\,\frac{\overline{\varphi'(\sigma)}}{\sigma-\zeta}\,d\sigma = \frac{1}{2\pi i}\oint_\gamma \frac{f}{\sigma-\zeta}\,d\sigma \quad , \tag{8}$$

$$\psi(\zeta) = \frac{1}{2\pi i}\oint_\gamma \frac{\bar f}{\sigma-\zeta}\,d\sigma - \frac{1}{2\pi i}\oint_\gamma \frac{\overline{\omega(\sigma)}}{\omega'(\sigma)}\,\frac{\varphi'(\sigma)}{\sigma-\zeta}\,d\sigma - \overline{\varphi(0)} \quad , \tag{9}$$

with γ being the unit circle $\zeta=\sigma=e^{i\vartheta}$, and

$$f = i\int_{z_0}^{z}(p_x + i\,p_y)\,ds = \int_{\sigma_0}^{\sigma} p\,|\omega'(\sigma)|\frac{d\sigma}{\sigma} \quad . \tag{10}$$

Since $\omega(\zeta)$ is a polynomial, $\omega(\sigma)/\overline{\omega'(\sigma)}$ will be a rational function:

$$\frac{\omega(\sigma)}{\overline{\omega'(\sigma)}} = \frac{c_1\sigma + \cdots + c_n\sigma^n}{\bar c_1 + \cdots + n\bar c_n\sigma^{-n+1}} = \sigma^n\frac{c_1 + \cdots + c_n\sigma^{n-1}}{\bar c_1\sigma^{n-1} + \cdots + n\bar c_n} \quad . \tag{11}$$

If we expand the above equation into a Laurent series, the highest power of σ is only n. So

$$\frac{\omega(\sigma)}{\overline{\omega'(\sigma)}}\,\frac{1}{\sigma-\zeta} = \left(b_n\sigma^n + \cdots + b_1\sigma + b_0 + \cdots\right)\left(\frac{1}{\sigma} + \frac{\zeta}{\sigma^2} + \frac{\zeta^2}{\sigma^3} + \cdots\right) \quad . \tag{12}$$

As we will see below the terms of the series are only to be taken into account up to b_0. Since $\bar\varphi'(1/\zeta)$ is analytical in $|\zeta|>1$ and $\bar\varphi'(0)$ is finite, we have

$$\oint_\gamma \frac{\overline{\varphi'(\sigma)}}{\sigma^k}\,d\sigma = 0 \qquad (k=2,\ 3,\ \cdots) \ . \tag{13}$$

Running out the parentheses of eq. (12) and inserting the result into eq. (8), we get an integral equation with a degenerated kernel:

$$\varphi(\zeta) + \frac{1}{2\pi i}\oint_\gamma \sum_{\nu=0}^{n}\zeta^\nu a_\nu(\sigma)\overline{\varphi'(\sigma)}\,d\sigma = \frac{1}{2\pi i}\oint_\gamma \frac{f}{\sigma-\zeta}\,d\sigma \quad . \tag{14}$$

Here

$$a_\nu(\sigma) = \sum_{k=\nu}^{n} b_k\sigma^{k-\nu-1} \qquad (\nu = 0, 1, \cdots, n) \quad . \tag{15}$$

Denoting

$$K_\nu = \frac{1}{2\pi i}\oint_\gamma a_\nu(\sigma)\overline{\varphi'(\sigma)}\,d\sigma \qquad (\nu = 0, 1, \cdots, n) \quad , \tag{16}$$

then differentiating eq. (14) with respect to ζ and integrating by parts on the right hand side of the equation, and bearing in mind that p is a self-equilibrated force system, we have

$$\varphi'(\zeta)+\sum_{\nu=1}^{n}\nu K_\nu \zeta^{\nu-1}=\frac{1}{2\pi i}\oint_\gamma \frac{p|\omega'(\sigma)|}{\sigma(\sigma-\zeta)}d\sigma \quad . \tag{17}$$

Since

$$\frac{1}{2\pi i}\oint_\gamma \frac{\omega(\sigma)}{\overline{\omega'(\sigma)}}\frac{\overline{\varphi'(\sigma)}}{\sigma-\zeta}d\sigma=\sum_{\nu=0}^{n}K_\nu\zeta^\nu \ , \ (|\zeta|<1) \quad , \tag{18}$$

the following equation for $\psi'(\zeta)$ can be derived from eq.(9):

$$\psi'(\zeta)=\frac{-1}{2\pi i}\oint_\gamma \frac{\overline{p(\sigma)}\,|\omega'(\sigma)|}{\sigma(\sigma-\zeta)}d\sigma-\frac{d}{d\zeta}\left[\frac{\bar{\omega}\left(\frac{1}{\zeta}\right)}{\omega'(\zeta)}\varphi'(\zeta)\right]-\sum_{\nu=1}^{n}\frac{\nu\bar{K}_\nu}{\zeta^{\nu+1}} \quad . \tag{19}$$

There are n complex constants $K_1,\cdots,K_n$ to be determined in eq.(17) and (19). In order to form the kernel functions, these constants should also be represented by the integration of the distributed traction p. Let $\zeta\to t$ ($|t|=1$), according to the Plemelj formula, eq.(17) is written as follows:

$$\varphi'(t)+\sum_{\nu=1}^{n}\nu K_\nu t^{\nu-1}=\frac{1}{2\pi i}\oint_\gamma \frac{p|\omega'(\sigma)|}{\sigma(\sigma-t)}d\sigma+\frac{p(t)\,|\omega'(t)|}{2t} \quad . \tag{20}$$

In order to find a system of equations for K_j (j = 1,...n) we first take the conjugate complex of the above equation, and then multiply it by $a_j(t)/(2\pi i)$ and integrate over t. Then from the second term of eq. (20) we obtain

$$\sum_{\nu=1}^{n}\left(\frac{1}{2\pi i}\oint_\gamma \frac{a_j(t)}{t^{\nu-1}}dt\right)\nu\bar{K}_\nu=\sum_{\nu=1}^{n}\left(\frac{1}{2\pi i}\oint_\gamma \frac{\sum_{m=j}^{n}b_m t^{m-j-1}}{t^{\nu-1}}dt\right)\nu\bar{K}_\nu$$

$$=\sum_{m=j}^{n}(m-j+1)\,b_m\bar{K}_{m-j+1} \quad . \tag{21}$$

The first term on the right hand side yields

$$\frac{1}{2\pi i}\oint_\gamma \frac{t\,a_j(t)}{\sigma-t}dt=-\frac{1}{2}\sigma a_j(\sigma) \quad . \tag{22}$$

Then together with eq. (16) we finally get a set of equations for K_j:

$$K_j+\sum_{m=j}^{n}(m-j+1)\,b_m\bar{K}_{m-j+1}=\frac{1}{2\pi i}\oint_\gamma \overline{p(\sigma)}\,|\omega'(\sigma)|\,\sigma\, a_j(\sigma)\,d\sigma \ , \quad (j=1,\cdots,n) \quad . \tag{23}$$

After separating the real and imaginary parts of eq. (23), we will get $2n$ equations with which $2n$ real constants are represented by the integration of p. Then the stress can be computed from $\varphi'(\zeta)$, $\varphi''(\zeta)$, $\psi'(\zeta)$ which are also represented by the integration of p, and the influence coefficients can be calculated.

We must see that one of the $2n$ real constants in eq.(23) remains undetermined; because of eq.(6), any imaginary constant added to $\varphi'(\zeta)/\omega'(\zeta)$ does not affect the stress. In order to close the system of equations, we may put $Im(K_1)=0$.

Example: A simple example for the polynomial mapping is

$$z=\omega(\zeta)=R(\zeta+m\zeta^2) \quad (R \text{ real and positive, } 0\le m\le 1/2) \quad . \tag{24}$$

Therefore

$$\frac{\omega(\sigma)}{\overline{\omega'(\sigma)}}=\frac{\sigma+m\sigma^2}{1+2m\sigma^{-1}}=m\sigma^2+(1-2m^2)\sigma+(4m^3-2m)+\cdots \quad . \tag{25}$$

Consequently we have

$$b_1=1-2m^2, \qquad b_2=m \quad , \tag{26}$$

$$a_1=m+(1-2m^2)/\sigma, \qquad a_2=m/\sigma \quad , \tag{27}$$

$$f_1=\frac{R}{2\pi i}\oint_\gamma \overline{p(\sigma)}\,|1+2m\sigma|\,(m\sigma+1-2m^2)\,d\sigma \quad , \tag{28)a}$$

$$f_2=\frac{R}{2\pi i}\oint_\gamma \overline{p(\sigma)}\,|1+2m\sigma|\,m\,d\sigma \tag{28)b}$$

Substituting eq. (26 - 28b) into eq. (23) yields

$$\begin{pmatrix} K_1 \\ K_2 \end{pmatrix}+\begin{pmatrix} 1-2m^2 & 2m \\ m & 0 \end{pmatrix}\begin{pmatrix} \overline{K}_1 \\ \overline{K}_2 \end{pmatrix}=\begin{pmatrix} f_1 \\ f_2 \end{pmatrix} \quad . \tag{29}$$

The real and imaginary parts of the above equations are

$$\begin{pmatrix} 2-2m^2 & 2m \\ m & 1 \end{pmatrix}\begin{pmatrix} K_{1r} \\ K_{2r} \end{pmatrix}=\begin{pmatrix} f_{1r} \\ f_{2r} \end{pmatrix} \quad , \tag{30}$$

$$\begin{pmatrix} 2m^2 & -2m \\ -m & 1 \end{pmatrix}\begin{pmatrix} K_{1i} \\ K_{2i} \end{pmatrix}=\begin{pmatrix} f_{1i} \\ f_{2i} \end{pmatrix} \quad . \tag{31}$$

K_{1r} and K_{2r} can be solved from eq. (30). However, since the rank of the matrix in eq. (31) is 1, K_{1i} and K_{2i} may have multiple solutions. Eq. (31) can be rewritten as

$$\begin{pmatrix} 0 & 0 \\ -m & 1 \end{pmatrix}\begin{pmatrix} K_{1i} \\ K_{2i} \end{pmatrix}=\begin{pmatrix} f_{1i}+2mf_{2i} \\ f_{2i} \end{pmatrix} \quad . \tag{31)'}$$

It is not difficult to prove that the first equation coincides with the equilibrium of the total moment of the force distribution acting along the boundary:

$$f_{1i}+2mf_{2i}=Im[f_1+2mf_2]=\frac{1}{2\pi R}\oint_L (yp_x-xp_y)\,ds \quad . \tag{32}$$

The results of (K_{1i},K_{2i}) can be taken as $(0,f_{2i})$ or $(-f_{2i}/m,\ 0)$.

SIMPLE RATIONAL FRACTIONAL MAPPING FUNCTION

When the covering body A_k is an infinite region with a hole, the mapping $\omega(\zeta)$ is always a rational fractional function. When $\omega(\zeta)$ can be rewritten in the simple form

$$z=\omega(\zeta)=\frac{c}{\zeta}+c_1\zeta+c_2\zeta^2+\cdots+c_n\zeta^n \tag{33}$$

then the highest power in the Laurent series of $\omega(\sigma)/\overline{\omega'(\sigma)}$ turns out as n-2. If $n>2$, the method still can be used to get the solution. If n=1, 2, we don't need to introduce the constants K_j since the second term in eq. (8) will vanish.

Consider the mapping of an elliptic domain onto the interior of the unit circle, where

$$z=\omega(\zeta)=R\left(\zeta+\frac{m}{\zeta}\right) \quad . \tag{34}$$

Then the region outside of the elliptic hole corresponds in the ζ-plane to the exterior of the unit circle. The boundary condition in this case is, if $\omega(\sigma)/\overline{\omega'(\sigma)}$ is expressed in terms of σ,

$$\varphi(\sigma)+\frac{\sigma^2+m}{\sigma(1-m\sigma^2)}\overline{\varphi'(\sigma)}+\overline{\psi(\sigma)}=f(\sigma) \quad , \tag{35}$$

where $f(\sigma)$ differs from that in eq.(10) by the sign:

$$f=i\int_{z_0}^{z}(p_x+ip_y)\,ds=-\int_{\sigma_0}^{\sigma}p|\omega'(\sigma)|\frac{d\sigma}{\sigma} \tag{36}$$

since in the image plane the sense of increasing ϑ ($e^{i\vartheta}=\sigma$) is opposite to that of the boundary. Taking Cauchy's integral over the unit circle, and noticing that $\varphi'(\infty)=0$, one can derive directly the equation

$$-\varphi(\zeta)=\frac{1}{2\pi i}\oint_{\gamma}\frac{f(\sigma)}{\sigma-\zeta}\,d\sigma \quad . \tag{37}$$

Here the positive direction of γ is counter-clockwise. It is opposite to that of the boundary, so a change of sign is produced again in the above equation. Assume that p is in equilibrium itself. Differentiating eq. (37) with respect to ζ and then integrating by part on the right hand side, it follows,

$$\varphi'(\zeta)=\frac{1}{2\pi i}\oint_{\gamma}\frac{p|\omega'(\sigma)|}{\sigma(\sigma-\zeta)}\,d\sigma \quad . \tag{38}$$

In the same way, the function $\psi'(\zeta)$ may be developed from the conjugate complex of eq. (35) as follows

$$\psi'(\zeta)=-\frac{\zeta(1+m\zeta^2)}{\zeta^2-m}\left[\varphi''(\zeta)+\left(\frac{1}{\zeta}+\frac{2m\zeta}{1+m\zeta^2}-\frac{2\zeta}{\zeta^2-m}\right)\varphi'(\zeta)\right]-\frac{1}{2\pi i}\oint_{\gamma}\frac{\bar{p}|\omega'(\sigma)|}{\sigma(\sigma-\zeta)}\,d\sigma \,. \tag{39}$$

The above formulas can be used to solve the elliptic hole problem in an infinite plate as well as an example with an elliptic notch in a semi-infinite plate , see Fig. 2 . There, of course the unknown function $p=p_x+i\,p_y$ is only defined in $0\le\vartheta\le\pi/2$.

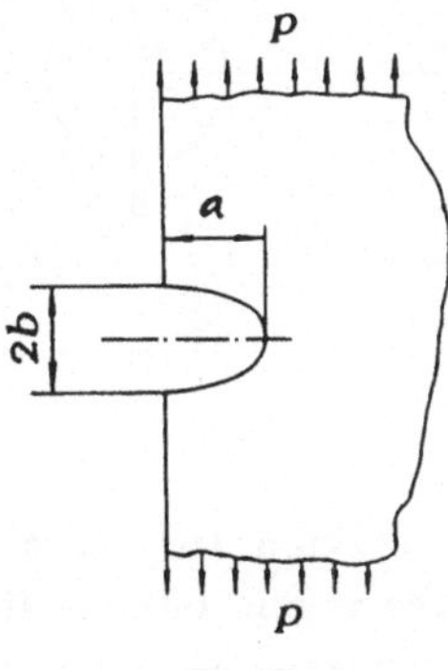

Fig. 2

Table.1 The Stress Concentration in the Root of the Elliptic Notch

b/a	σ_y/p
0.2	11.7183
0.4	6.2546
0.6	4.4595
0.8	3.5726
1.0	3.0456

ONE EXAMPLE FOR A GENERAL RATIONAL FRACTIONAL MAPPING FUNCTION

There will be several poles in the second term on eq.(8) when the mapping is a general rational fractional function, say the ratio of two general polynomials. Then the Cauchy integration must be dealt with by the theorem of residues. As will be seen, in this case, several constants K_j will still be produced. However, since most poles are distributed over a finite region, they can be determined easily.

As one example we are going to present some techniques by dealing with a lip crack problem, in which the mapping function is a kind of general rational fractional function. Suppose the covering domain to be an infinite plane with a lip crack, the length of which is $2a$, and the height is $2h$, see Fig. 3. The boundary of the lip crack in the z-plane can be mapped onto an ellipse in an intermediate image plane (t-plane) by the following mapping

$$z=\frac{a}{2}\left(t+\frac{1}{t}\right) \quad . \tag{40}$$

The long and short seminaxes of the ellipse are T and 1, respectively, where

$$T:=\frac{h}{a}+\sqrt{\left(\frac{h}{a}\right)^2+1} \quad . \tag{41}$$

Then the ellipse can be mapped onto the unit circle in the ζ-plane by the already mentioned mapping

$$t=R\left(\zeta-\frac{m^2}{\zeta}\right) \quad , \tag{42}$$

where now

$$R:=\frac{1}{2}(T+1), \qquad m^2:=\frac{T-1}{T+1} \quad . \tag{43}$$

Combining eq.(40) and eq.(42) we can write the mapping function $\omega(\zeta)$ as

$$z=\omega(\zeta)=\frac{a}{2}\left[R\zeta-\frac{Rm^2}{\zeta}+\frac{\zeta}{R(\zeta^2-m^2)}\right] \quad . \tag{44}$$

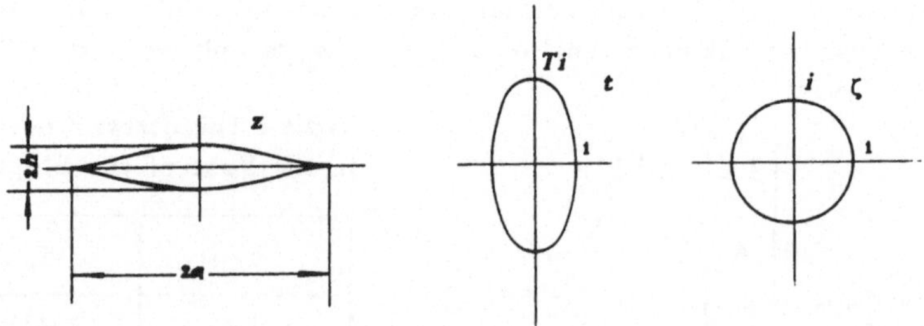

Fig.3

Eq. (44) appears as a typical rational fractional function for mappings of the region outside the unit circle in the ζ-plane onto the region outside the lip crack in the z-plane where both real axes in the two planes are corresponding, too.

The inverse form of eq. (44) can be deduced also from eq. (40) and eq. (42) as

$$\zeta=\frac{t}{2R}+\sqrt{\left(\frac{t}{2R}\right)^2+m^2}\;,\qquad t=\frac{z}{a}+\sqrt{\left(\frac{z}{a}\right)^2-1}\quad. \tag{45}$$

Some constants K_j like those in eq. (23) must be introduced in dealing with the problem related to the mapping eq. (44). For convenience, the function $\chi_1(z)$, or $\chi(\zeta)$ will be used, where $\psi(\zeta)=\chi_1'(z)=\chi'(\zeta)/\omega'(\zeta)$. So the boundary condition is

$$\varphi(\sigma)+\frac{\omega(\sigma)}{\overline{\omega'(\sigma)}}\,\overline{\varphi'(\sigma)}+\frac{1}{\overline{\omega'(\sigma)}}\,\overline{\chi'(\sigma)}=f(\sigma)\quad. \tag{46}$$

Since the problem is in the outside region, f has the same form as in eq. (36). Denote

$$\Omega(\sigma)=\frac{\omega(\sigma)}{\overline{\omega'(\sigma)}}=\frac{\sigma-\dfrac{m^2}{\sigma}+\dfrac{\sigma}{R^2(\sigma^2-m^2)}}{(1+m^2\sigma^2)\left[1-\dfrac{\sigma^2}{R^2(1-m^2\sigma^2)^2}\right]}\;, \tag{47}$$

$$\Omega_0(\sigma)=\frac{1}{\overline{\omega'(\sigma)}}=\frac{1}{\dfrac{Ra}{2}(1+m^2\sigma^2)\left[1-\dfrac{\sigma^2}{R^2(1-m^2\sigma^2)^2}\right]}\quad. \tag{48}$$

Then eq. (46) can be written as

$$\varphi(\sigma)+\Omega(\sigma)\overline{\varphi'(\sigma)}+\Omega_0(\sigma)\overline{\chi'(\sigma)}=f(\sigma)\quad. \tag{49}$$

Taking w as the integration variable in the ζ-plane it is not difficult to prove that there are five poles of order 1 in the region $|w|\le 1$ for the function $\Omega(w)$, namely $w=0,\pm m,\ \pm 1$, and there are two poles of order 1 for $\Omega_0(w)$: $w=\pm 1$. Using the theorem of residues, noticing that the functions $\bar{\varphi}'(1/w)$ and $\bar{\chi}'(1/w)$ are analytical on $|w|\le 1$, one can develop the following result,

$$-\varphi(\zeta)+\frac{b}{m-\zeta}\,\bar{\varphi}'\left(\frac{1}{m}\right)-\frac{b}{m+\zeta}\,\bar{\varphi}'\left(\frac{-1}{m}\right)-\frac{1}{2T^2}\,\frac{\bar{\varphi}'(1)}{1-\zeta}+\frac{1}{2T^2}\,\frac{\bar{\varphi}'(-1)}{1+\zeta}$$

$$-\frac{1}{2aT^2}\,\frac{\bar{\chi}'(1)}{1-\zeta}-\frac{1}{2aT^2}\,\frac{\bar{\chi}'(-1)}{1+\zeta}=\frac{1}{2\pi i}\oint_\gamma\frac{f(\sigma)}{\sigma-\zeta}\,d\sigma\;, \tag{50}$$

where

$$b:=\lim_{w\to m}(w-m)\Omega(w)=\frac{4T^2}{(T^2+1)(3T^2+1)}\quad. \tag{51}$$

The poles $w=\pm 1$ being on the circle γ produce the factor 1/2 in the related terms. Since the boundary in the ζ-plane is smooth, $\varphi(\zeta)$ should be continuous on the boundary. Therefore

$$-\frac{\bar{\varphi}'(1)}{2T^2}-\frac{\bar{\chi}'(1)}{2aT^2}=0,\qquad \frac{\bar{\varphi}'(-1)}{2T^2}-\frac{\bar{\chi}'(-1)}{2aT^2}=0\quad. \tag{52}$$

Denote

$$K_1:=\varphi'(1/m)\;,\qquad K_2:=\varphi'(-1/m)\quad, \tag{53}$$

and rewrite eq. (50) to find

$$\varphi(\zeta)+\frac{b}{\zeta-m}\overline{K}_1+\frac{b}{\zeta+m}\overline{K}_2=-\frac{1}{2\pi i}\oint_{\gamma}\frac{f(\sigma)}{\sigma-\zeta}d\sigma \quad . \tag{54}$$

Then

$$\varphi'(\zeta)=\frac{b}{(\zeta-m)^2}\overline{K}_1+\frac{b}{(\zeta+m)^2}\overline{K}_2+\frac{1}{2\pi i}\oint_{\gamma}\frac{F|\omega'(\sigma)|}{\sigma(\sigma-\zeta)}d\sigma \quad . \tag{55}$$

The function $\chi'(\zeta)$ (or $\psi(\zeta)$) can be developed also by the theorem of residue, here only the result is given:

$$\psi'(\zeta)=-\frac{1}{2\pi i}\oint_{\gamma}\frac{\overline{F}|\omega'(\sigma)|}{\sigma(\sigma-\zeta)}d\sigma+\frac{b}{m^2(\zeta-1/m)^2}K_1+\frac{b}{m^2(\zeta+1/m)^2}K_2$$

$$-\Gamma'(\zeta)\varphi'(\zeta)-\Gamma(\zeta)\varphi''(\zeta) \quad , \tag{56}$$

$$\Gamma(\zeta)=\overline{\Omega}\left(\frac{1}{\zeta}\right)=\frac{\overline{\omega}(1/\zeta)}{\omega'(\zeta)} \quad . \tag{57}$$

The final work, as presented for $\omega(\zeta)$ being a polynomial, is to determine the coefficients K_1 and K_2. Let $\zeta=\pm 1/m$ in eq. (55) , we get

$$\left.\begin{aligned} K_1&=\frac{b}{(m-\frac{1}{m})^2}\overline{K}_1+\frac{b}{(m+\frac{1}{m})^2}\overline{K}_2+\frac{1}{2\pi i}\oint_{\gamma}\frac{F|\omega'(\sigma)|}{\sigma(\sigma-1/m)}d\sigma \; , \\ K_2&=\frac{b}{(m+\frac{1}{m})^2}\overline{K}_1+\frac{b}{(m-\frac{1}{m})^2}\overline{K}_2+\frac{1}{2\pi i}\oint_{\gamma}\frac{F|\omega'(\sigma)|}{\sigma(\sigma+1/m)}d\sigma \; . \end{aligned}\right\} \tag{58}$$

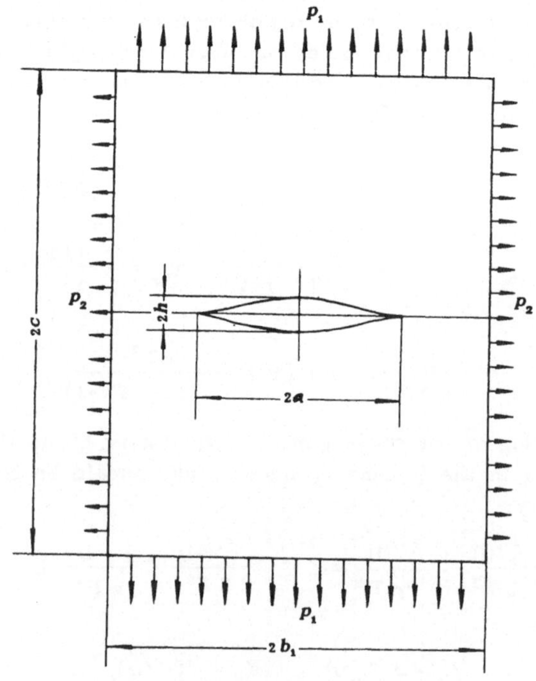

Fig. 4

Since in eq. (58) K_1 and K_2 are represented by integrations of p over the boundary, the influence coefficient will be formed. After superposition in the physical plane, it remains to solve a system of linear equations which corresponds to the system of integral equations. Finally the stress intensity factors K_I (mode I) and K_{II} (mode II) at the tip point of the lip crack can be calculated

$$K_I - iK_{II} = 2 \lim_{\zeta\to 1} \sqrt{2\pi}\,[\omega(\zeta)-\omega(1)]^{1/2} \frac{\varphi'(\zeta)}{\omega'(\zeta)} = 2\sqrt{\pi}\,\frac{\varphi'(1)}{\sqrt{\omega''(1)}}$$

$$= \frac{2\sqrt{\pi}}{\sqrt{aT}} \left[\frac{1}{2\pi} \int_{-\pi}^{\pi} \frac{F|\omega'(\sigma)|}{\sigma - 1}\, d\vartheta + \frac{b}{(1-m)^2}\,\bar{K}_1 + \frac{b}{(1+m)^2}\,\bar{K}_2 \right] . \qquad (59)$$

Example: According to the above formulas we have calculated a rectangular plate with a central lip crack, see Fig. 4, where the covering domains for four straight lines are taken as the semi-infinite planes. The calculations are taken under two load conditions, one being the vertical tension p_1 and another the transverse tension p_2. The results, denoted by $F_1 := K_I/(p_1\sqrt{\pi a})$, $F_2 := K_I/(p_2\sqrt{\pi a})$, are shown in Fig. 5-6. It is interesting that $F_2 \le 0$ in most cases, which means that the crack still tends towards opening while the plate is pressed transversely.

CONCLUSIONS AND ACKNOWLEDGEMENT

We have discussed the covering domain method for plane problems in static elasticity within the framework of nonsingular Fredholm integral equations of the second kind, and the technique to form an approach for the influence coefficients by conformal mapping. For the mapping $\omega(\zeta)$, the polynomial and rational fractional functions have been considered. Three examples were given. However, the method described here can be applied to the case where $\omega(\zeta)$ has a more complicated form. So there is a wealth of applications.

The authors wish to express their gratitude to the Alexander von Humboldt Foundation for the research fellowship of Xiao Lin, and to the East China Institute of Technology for providing the opportunity to realize this cooperation.

REFERENCES

[1] S.G. Mikhlin: Integral Equations and Their Applications to Certain Problems in Mechanics, Mathematical Physics and Technology. Pergamon Press, London -New York-Paris-Los Angeles, 1957, p239-242.

[2] R.J. Hartranft and G.C. Sih: Alternating Method Applied to Edge and Surface Crack Problems. In Mechanics of Fracture I, ed. by G.C. Sih, Noordhoff International Publishing, Leyden (1973).

[3] X. Lin and L.G. Wang: General Expressions of Fredholm Integral Equations Method on Elastic Bodies with Cracks and its Discussion. Proc. Int. Conf. on Fracture and Fracture Mechanics, Shanghai, April, 1987, pp107-110.

[4] X. Lin and L.G. Wang: Some Approaches to Form Interacting Functions in the Fredholm Integral Equations Method of Elastic Mechanics, Acta Mechanica Sinica, Vol. 19, pp417-425, (1987), (in Chinese)

[5] N.I. Muskhelishvili: Some Basic Problems of Mathematical Theory of Elasticity, Noordhoff Groningen, Holland, (1953).
[6] Y.Z. Chen: Solution of Multiple Crack Problems of a Circular Plate or an Infinite Plate Containing a Circular Hole by Using Fredholm Integral Equation Approach. Int. J. of Fracture, Vol.25, No. 1 (1984).

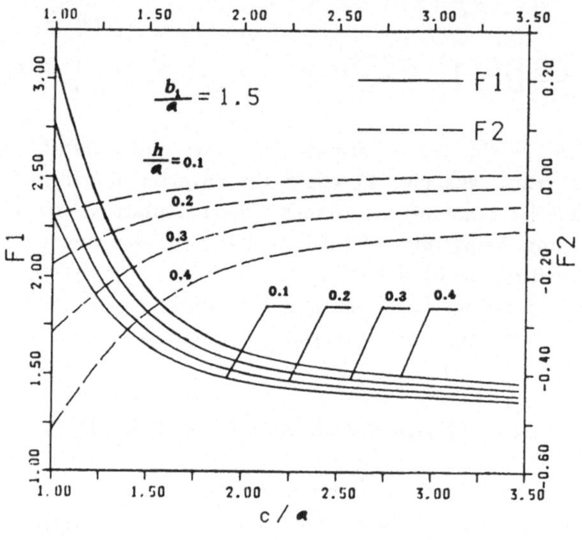

Fig.5

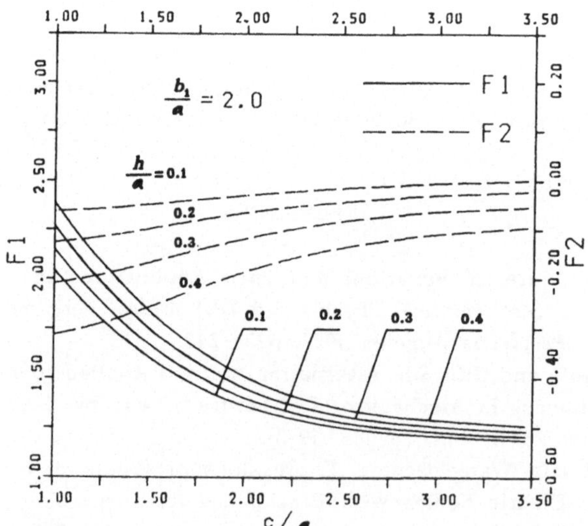

Fig.6

ON QUADRATURE METHODS OF GAUSS TYPE FOR SINGULAR INTEGRAL EQUATIONS AND THE AIRFOIL EQUATION

G. Mastroianni* and S. Prössdorf**

*Istituto per Applicazioni della Matematica, C.N.R., Via P. Castellino 111, 80131 Napoli, Italy
**Karl-Weierstraß-Institut für Mathematik, Mohrenstraße 39, D - O-1086 Berlin, Germany

SUMMARY

The weighted mean convergence of a Gauss type quadrature method for solving singular integral equations over the arc $(-1,1)$ with Cauchy kernel and a not necessarily regular perturbation kernel is studied. Moreover, error estimates in uniform norms are given. Finally the method is applied to the numerical solution of Prandtl's airfoil equation.

INTRODUCTION

Many problems in aerodynamics and elasticity lead to a singular integral equation with Cauchy kernel of the form

$$a\,u(x) + \frac{b}{\pi}\int_{-1}^{1}\frac{u(t)}{t-x}\,dt + \int_{-1}^{1}k(x,t)u(t)\,dt = f(x) \tag{1}$$

on the interval (-1,1) (see e.g. [1], [16], [19]). The first integral in (1) is to be interpreted as the Cauchy principal value. Hereby a, b and f are given Hölder continuous functions, and k is a given smooth or weakly singular kernel function.

The problem we are interested in is to find an approximation to the unknown solution u by using projection methods (like collocation or Galerkin schemes) or quadrature procedures with orthogonal polynomials as trial functions. There is a considerable literature on this subject in the case of regular kernel k (see, e.g., the surveys [9], [6]–[8], [12], [22], [23], [13]–[15], [24] and the references given by the same authors). In the most of these papers the following strategy is employed. For given functions a and b, one introduces two sets of orthogonal polynomials which are denoted by $\{p_n\}$ and $\{q_n\}$, where $Dp_n = q_{n-\chi}$ with D being the dominant part of Eq. (1) and χ the index of D (see Section 1). For a given value of n, we use Gauss type quadrature rules based on the zeros of p_n and collocate at the zeros of $q_{n-\chi}$. In the case of constant coefficients a and b, the polynomials p_n and q_n turn out to be Jacobi polynomials corresponding to the weight functions $\omega(t) = \omega^{\alpha,\beta}(t) = (1-t)^{\alpha}(1+t)^{\alpha}$, $-1 < \alpha, \beta < 1$, and $1/\omega(t)$, respectively.

If, e.g., $k \in C^{r+\lambda}([-1,1]^2)$ and $f \in C^{r+\lambda}[-1,1]$, where r is a non-negative integer and $0 < \lambda \leq 1$, and $|\alpha| = |\beta| = \frac{1}{2}$, $\chi = 0$ or $\chi = 1$, then in [14] the corresponding quadrature method is proved to converge in the weighted space $L^2_\omega(-1,1)$ with the rate $O(n^{-r-\lambda})$ as $n \to \infty$.

In the present paper we prove that for real constant coefficients a and b the quadrature method converges in the space $L^2_\omega(-1,1)$ with the error bound $O(n^{-r-\lambda}\log n)$, provided only that k is a kernel of the form $k(x,t) = [h(x,t) - h(x,x)]/(t-x)$, where $h \in C^{r+\lambda}([-1,1]^2)$, $f \in C^{r+\lambda}[-1,1]$, and α, β are arbitrary numbers satisfying $-1 < \alpha, \beta < 1$ and $\chi = 0$ or $\chi = 1$. Moreover, error estimates in uniform norms are given. The crucial point in our analysis are bounds of the quadrature error for the perturbation kernel k which are founded on thorough estimates for the distance between the zeros of the orthogonal polynomials p_n and q_n. Further ideas of Junghanns and Silbermann [14] and of Elliott [5]-[8] are used. The results apply to the numerical solution of Prandtl's equation in wing theory. For the case of variable coefficients a and b as well as for the detailed proofs the reader is referred to our paper [25].

SINGULAR INTEGRAL EQUATIONS WITH CONSTANT COEFFICIENTS

Consider the singular integral equation (1), where a and b are given real constants such that $a^2 + b^2 > 0$, and k is a regular or weakly singular kernel function,

$$k(x,t) = \frac{h(x,t) - h(x,x)}{t-x} \tag{1'}$$

with $h \in C^\lambda([-1,1]^2)$, $0 < \lambda \leq 1$.

Notice that if k is a weakly singular kernel of the form

$$k(x,t) = \frac{m(x,t)}{|x-t|^\mu}, \quad 0 \leq \mu < 1,$$

with $m \in C^\nu([-1,1]^2)$, then the representation (1') holds, where $h(x,t) = (t-x)|t-x|^{-\mu}m(x,t) \in C^\lambda([-1,1]^2)$ with $\lambda = \min(1-\mu, \nu)$ (see e.g. [19]).

The quadrature method under consideration, like most of the polynomial approximation methods for solving equation (1), is essentially based on the well-known relation (see e.g. [9], [5]-[6], [13]-[15])

$$D(p_n^{(\alpha,\beta)}) = -\frac{2^{-\chi}b}{\sin(\pi\alpha_0)} p_{n-\chi}^{(-\alpha,-\beta)}, \tag{2}$$

where

$$(Dv)(t) = a\,\omega(x)v(x) + \frac{b}{\pi}\int_{-1}^{1} \frac{\omega(t)v(t)}{t-x}\,dt, \quad -1 < x < 1, \tag{3}$$

is the dominant part of Eq. (1), and $\{p_n^{\alpha,\beta}\}_{n\in\mathbf{N}}$ is the sequence of orthonormal Jacobi polynomials respect to the weight function $\omega(x) = \omega^{(\alpha,\beta)}(x) = (1-x)^\alpha(1+x)^\beta$, that is, $p_n^{(\alpha,\beta)}$ is a polynomial of degree n with positive leading coefficient and $\int_{-1}^1 p_n^{(\alpha,\beta)} p_m^{(\alpha,\beta)}\omega = \delta_{n,m}$ ($p_{n-\chi}^{(-\alpha,-\beta)} \equiv 0$ for $n-\chi<0$). The numbers α, β and χ involved are defined as follows:

$$\begin{aligned} a+ib &= (a^2+b^2)^{1/2}e^{i\pi\alpha_0}, \quad 0<|\alpha_0|<1, \\ -1<\alpha: &= \mu-\alpha_0,\ \beta := \nu+\alpha_0<1, \end{aligned}$$

μ and ν are integers, $\chi := -(\alpha+\beta) = -(\mu+\nu)$.

Denote by $L_\omega^2 = L_\omega^2(-1,1)$ the Hilbert space of all complex-valued functions on $(-1,1)$ which are square integrable with respect to the weight ω. The space L_ω^2 is equipped with the scalar product

$$(u,v)_\omega = \frac{1}{\pi}\int_{-1}^1 u(t)\overline{v(t)}\omega(t)\,dt$$

and the norm $\|u\|_\omega = [(u,u)_\omega]^{1/2}$. As an immediate consequence of (2), the operator D defined by (3) and acting from L_ω^2 to $L_{1/\omega}^2$ is Fredholm with index χ (see also [11], [18]). Moreover, D is invertible if $\chi = 0$, invertible from the right if $\chi = 1$ (in this case $D(\mathbf{P}_{n-1}) = \mathbf{P}_{n-2}$ with $\mathbf{P}_{n-1}$ being the set of all real polynomials of degree $\leq n$) and invertible from the left if $\chi = -1$ (in which case $D(\mathbf{P}_{n-1}) + \mathbf{C} = \mathbf{P}_n$).

Let $t_k = t_{m,k}$, $k = 1, ..., m$, and $x_j = x_{m,j}$, $j = 1, ..., m-\chi$, be the zeros of the polynomials $p_m^{(\alpha,\beta)}$ and $p_{m-\chi}^{(-\alpha,-\beta)}$, respectively, i.e.

$$p_m^{(\alpha,\beta)}(t_k) = 0, \quad p_{m-\chi}^{(-\alpha,-\beta)}(x_j) = 0.$$

Choose now a Gauss-type quadrature formula with the weight ω and the nodes t_k such that

$$\int_{-1}^1 g(t)\omega(t)\,dt = \sum_{k=1}^m \lambda_{m,k} g(t_k), \quad g \in \mathbf{P}_{2m-1}, \tag{4}$$

where $\lambda_{m,k} = \lambda_{m,k}(\omega)$ stand for the Christoffel numbers. Then

$$(Dv_m)(x_j) = \frac{b}{\pi}\sum_{k=1}^m \lambda_{m,k}\frac{v_m(t_k)}{t_k - x_j}, \quad j = 1, ..., m-\chi, \tag{5}$$

holds for any polynomial v_m of degree $m-1$ (see e.g. [15, Lemma 1.15]).

We consider the following quadrature method for the approximate solution of Eq. (1). Determine $\xi_k = \xi_{m,k}$, $k = 1, ..., m$, such that

$$\sum_{k=1}^{m} \lambda_{m,k} \left[\frac{b/\pi}{t_k - x_j} + k(x_j, t_k) \right] \xi_k = f(x_j), \quad j = 1, ..., m - \chi. \tag{6}$$

In the case $\chi = 1$ it is necessary to give an additional condition in order to define the solution of Eq. (1) uniquely, e.g.

$$\int_{-1}^{1} u(t)\, dt = 0 \tag{7}$$

which can be approximated by

$$\sum_{k=1}^{m} \lambda_{m,k} \xi_k = 0. \tag{8}$$

Given $0 < \lambda \leq 1$, an integer $r \geq 0$, and a subset $A \subseteq [-1,1]^2$, we denote by $C^{r+\lambda}(A)$ the class of all functions on A whose r–th derivatives belong to $Lip_\lambda(A)$.

The main results of the present paper are as follows.

Theorem 1. *Assume* $-1 < \alpha, \beta < 1$, $\chi = 0$ *or* $\chi = 1$, $k \in C^{r+\lambda}([-1,1]^2)$ *and* $f \in C^{r+\lambda}[-1,1]$, $r \geq 1$, $0 < \lambda \leq 1$. *If the problem (1) (for* $\chi = 0$*) or (1), (7) (for* $\chi = 1$*) has a unique solution* $u = \omega v$, $v \in L^2_\omega$, *then the system of equations (6) or (6), (8), respectively, is uniquely solvable for all sufficiently large* m *and*

$$||v - v_m||_\omega = O\left(\frac{1}{m^{r+\lambda}}\right), \tag{9}$$

where

$$v_m(t) = \sum_{k=1}^{m} \frac{p_m^{(\alpha,\beta)}(t)}{(t - t_k) p_m^{(\alpha,\beta)\prime}(t_k)} \xi_k$$

is the Lagrange interpolation polynomial corresponding to the solution of (6) or (6), (8), respectively.

Theorem 2. *Assume* $\chi = 0$ *and* $|\alpha| \leq 1/2$ *or* $\chi = 1$ *and* $-1 < \alpha < 0$, $h \in C^{r+\lambda}([-1,1]^2)$ *(recall (1')) and* $f \in C^{r+\lambda}[-1,1]$, *where* $r \geq 0$ *and* $0 < \lambda \leq 1$. *Then the assertion of Theorem 1 remains true with the error bound*

$$||v - v_m||_\omega = O\left(\frac{\log m}{m^{r+\lambda}}\right). \tag{10}$$

Theorem 3. *Assume the hypotheses of Theorem 2 are satisfied. If* $r + \lambda > \gamma$, *where* $\gamma := \max(\alpha, \beta) + 1$, *then* $v \in C[-1,1]$ *and*

$$\max_{-1 \leq t \leq 1} |v(t) - v_m(t)| = O\left(\frac{\log m}{m^{r+\lambda-\gamma}}\right). \tag{11}$$

Further, if $r+\lambda > 1/2$, then v is continuous on any closed set $\Delta \subset (-1,1)$ and

$$\max_{\Delta} |v(t) - v_m(t)| = O\left(\frac{\log m}{m^{r+\lambda-1/2}}\right). \tag{12}$$

Remark. Theorem 1 and 3 (under additional assumptions) were proved in [14] for the case $|\alpha| = |\beta| = 1/2$.

Note that Theorem 3 guarantees, for example, the uniform convergence in $[-1,1]$ and estimate (11) in the following particular cases:

1) $r \geq 1$ and $\chi = 0$, if $r+\lambda > \gamma = 1+|\alpha|$,
2) $r \geq 1$ and $\chi = 1$, if $r+\lambda > \gamma = 1+\max(\alpha, -1-\alpha)$,
3) $r = 0$ and $\chi = 1$, if $\lambda > \gamma = 1+\max(\alpha, -1-\alpha)$.

In the following the symbol "C" stands for some positive constant taking a different value each time it is used. It will always be clear what variables and indices the constants are independent of. If A and B are two expressions depending on some variables then we write

$$A \sim B \text{ iff } |AB^{-1}| \leq C \quad \textit{and} \quad |A^{-1}B| \leq C$$

uniformly for the variables in consideration.

In order to prove the preceding theorems we need the following auxiliary results. In particular, we shall use the following lemma that can be found in [2].

Lemma 1 *(Jackson). For any function $f \in C^r([-1,1]^2)$, $r \geq 0$, and for any positive integer n, there exists an algebraic polynomial $P_n(x,y)$ of degree n in x and y separately such that*

$$|f(x,y) - P_n(x,y)| \leq C\, n^{-r} \Omega_r\left(f; \frac{1}{n}\right),$$

where

$$\Omega_r(f;\delta) = \max_{0 \leq i \leq r} \omega(f_i^{r-i};\delta), \quad f_i^{r-i} = \frac{\partial^r f}{\partial x^{r-i} \partial y^i}, \quad \delta > 0,$$

and

$$\omega(f_i^{r-i};\delta) = \max_{h_1+h_2 \leq \delta} |f_i^{r-i}(x+h_1, y+h_2) - f_i^{r-i}(x,y)|, \quad h_1, h_2 \geq 0.$$

Let $k(x,t) \in C^r([-1,1]^2)$. For any continuous function f we define the operator K by

$$(Kf)(x) = \int_{-1}^{1} k(x,t) f(t) \omega(t)\, dt. \tag{12'}$$

If $L_m^{(1)} g$ denotes the Lagrange polynomial interpolating the bounded function g on the zeros t_k, $k = 1, 2, ..., m$, of $p_m^{(\alpha,\beta)}$, then we set

$$\overline{G}_m(x) = \int_{-1}^{1} L_{m,t}^{(1)}\{k(x,t)v_m(t)\}\omega^{(\alpha,\beta)}(t)\,dt,$$

where $L_{m,t}^{(1)}$ is the interpolating operator $L_m^{(1)}$ acting on the function $k(x,t)v_m(t)$ with respect to the variable t. Obviously,

$$\overline{G}_m(x) = \sum_{k=1}^{m} \lambda_{m,k}(\omega^{(\alpha,\beta)})k(x,t_k)v_m(t_k).$$

Further, denoting by $L_{m-\chi}^{(2)}g$, $\chi = -(\alpha+\beta) \in \{0,1\}$, the Lagrange polynomial interpolating g at the zeros x_j, $j = 1,2,...,m-\chi$, of $p_{m-\chi}^{(-\alpha,-\beta)}$, for any polynomial $v_m \in \mathbf{P}_{m-1}$ we define the operator K_m by

$$\begin{aligned}(K_m v_m)(x) &= (L_{m-\chi}^{(2)}\overline{G}_m)(x) = \sum_{j=1}^{m-\chi} l_{m-\chi,j}(x)\overline{G}_m(x_j)\\ &= \sum_{j=1}^{m-\chi} l_{m-\chi,j}(x)\left[\sum_{k=1}^{m} \lambda_{m,k}(\omega^{(\alpha,\beta)})k(x_j,t_k)v_m(t_k)\right],\end{aligned}$$

where $l_{m-\chi,j}(x)$ are the fundamental Lagrange polynomials corresponding to the knots x_j, $j = 1,2,...,m-\chi$.

Notice that if $k(x,t)$ is a polynomial of degree $m-2$ in the variables x and t separately, then by well-known properties of the Gaussian rules and of the Lagrange polynomials, we have

$$(K_m v_m)(x) = \int_{-1}^{1} k(x,t)v_m(t)\omega^{(\alpha,\beta)}(t)\,dt = (Kv_m)(x),$$

or, that is the same as $(K - K_m)v_m = 0$.

Therefore, for any function $k(x,t)$ and for any polynomial Q of degree $m-2$ in the variables x and t separately, we get

$$\begin{aligned}[(K-K_m)v_m](x) &= \int_{-1}^{1}[k(x,t)-Q(x,t)]v_m(t)\omega^{(\alpha,\beta)}(t)\,dt\\ &- \sum_{j=1}^{m-\chi} l_{m-\chi,j}(x)\left\{\sum_{k=1}^{m}\lambda_{m,k}(\omega^{(\alpha,\beta)})[k(x_j,t_k)-Q(x_j,t_k)]v_m(t_k)\right\}.\end{aligned}$$

Finally, setting

$$||g||_\infty = \sup_{[-1,1]^2} |g(x,y)|,$$

we can state the following

Lemma 2. *([25]) Assume $k \in C^r([-1,1]^2)$, $r \geq 0$, and $\alpha, \beta \in (-1,1)$. Then*

$$||(K - K_m)v_m||_{\omega^{(-\alpha,-\beta)}} \leq C||r_m k||_\infty ||v_m||_{\omega^{(\alpha,\beta)}},$$

where $r_m k = k - P_m$ with P_m being the Jackson polynomial corresponding to the function k (see Lemma 1), and $C = 2\sqrt{2}[B(1+\alpha,1+\beta)B(1-\alpha,1-\beta)]^{1/2}$ with B the Euler function.

In order to state a lemma similar to the previous one with $k(x,t)$ replaced by $k(x,t) = [h(x,t) - h(x,x)]/(t-x)$, where $h \in C^r([-1,1]^2)$, we need some other preliminary results.

Lemma 3. *Let $\{p_m^{(\alpha,\beta)}\}$, $\{p_m^{(-\alpha,-\beta)}\}$, $\alpha, \beta \in (-1,1)$, be the sequences of Jacobi polynomials corresponding to the weights $\omega^{(\alpha,\beta)}$ and $\omega^{(-\alpha,-\beta)}$, respectively. Denote by $t_k = t_{m,k} = \cos \tau_{m,k}$, $k = 1,2,...,m$, and $x_j = x_{m,j} = \cos \theta_{m,j}$, $j = 1,2,...,m$, the zeros of $p_m^{(\alpha,\beta)}$ and $p_m^{(-\alpha,-\beta)}$, respectively.*

If $\alpha + \beta = 0$, then

$$\min_{j,k} |\tau_{m,k} - \theta_{m,j}| \sim m^{-1}.$$

Furthermore, if $\alpha + \beta = -1$, then

$$\min_{j,k} |\tau_{m+1,k} - \theta_{m,j}| \sim m^{-1}.$$

The proof of the previous lemma is due to the authors and it can be found in [17].

Lemma 4. *([25]) Assume $-(\alpha+\beta) = \chi = 0$ and $|\alpha| \leq 1/2$ or $\chi = 1$ with $-1 < \alpha < 0$, and let $h \in C^r([-1,1]^2)$. Then*

$$||(K - K_m)v_m||_{\omega^{(-\alpha,-\beta)}} \leq C||r_n h||_\infty ||v_m||_{\omega^{(\alpha,\beta)}} \log m,$$

where $v_m \in \mathbf{P}_{m-1}$, $r_n h = h - P_n$ with P_n being the Jackson polynomial of degree $n = [m/2] - 1$ corresponding to the function h, and the constant C is independent of m and h.

Proof of Theorems 1 and 2.
Eq. (1) can be written in the form

$$Dv + Kv = f, \tag{13}$$

where K is the operator defined by (12'). Assume first $\chi = 1$.

We recall that $L_m^{(1)} f$ denotes the Lagrange interpolation polynomial of degree $m-1$ with the nodes t_k, $k = 1,2,..,m$, and $L_{m-1}^{(2)} f$ denotes the interpolation polynomial of degree $m-2$ with the nodes x_j, $j = 1,2,...,m-1$. By virtue of (5) the system (6) is equivalent to the operator equation

$$Dv_m + K_m v_m = L_{m-1}^{(2)} f, \tag{14}$$

where

$$(K_m v_m)(x) = L_{m-1}^{(2)} \int_{-1}^{1} L_{m,t}^{(1)}\{k(x,t)v_m(t)\}\omega(t)\, dt,$$

and $L^{(1)}_{m,t}$ stands for the interpolation operator $L^{(1)}_m$ applied to the function $k(x,t)v_m(t)$ with respect to the variable t.

Since the solution of Eq. (13) is unique in $L^2_{\omega,0} := \{v \in L^2_\omega : (v,1)_\omega = 0\}$ and the operator $D : L^2_{\omega,0} \to L^2_{1/\omega}$ is invertible, we conclude that the operator $D + K : L^2_{\omega,0} \to L^2_{1/\omega}$ has a bounded inverse, because $K : L^2_{\omega,0} \to L^2_{1/\omega}$ is a compact operator (see, e.g., [18]).

Provided (14) has a solution v_m then

$$(D+K)v_m = (K - K_m)v_m + L^{(2)}_{m-1}f.$$

Thus

$$\begin{aligned} ||v_m||_\omega &\le ||(D+K)^{-1}|| \left[||(K-K_m)v_m||_{1/\omega} + ||L^{(2)}_{m-1}f||_{1/\omega}\right] \\ &\le ||(D+K)^{-1}|| \left[\epsilon_m ||v_m||_\omega + ||(D+K_m)v_m||_{1/\omega}\right], \end{aligned}$$

where $\epsilon_m = O(m^{-r-\lambda})$ as $m \to \infty$ in the case of Theorem 1 and $\epsilon_m = O(m^{-r-\lambda}\log m) \to 0$ in the case of Theorem 2 (see Lemmas 1, 2 and 4). Consequently, for all sufficiently large m, the estimate

$$C||v_m||_\omega \le ||(D+K_m)v_m||_{1/\omega} \tag{15}$$

holds with a positive constant $C \le ||(D+K)^{-1}||^{-1} - \epsilon_m$. Since $(D+K_m)v_m \in \mathrm{im}L^{(2)}_{m-1}$ for all $v_m \in \mathrm{im}L^{(1)}_m$, the estimate (15) implies the invertibility of the finite dimensional operator $D + K_m : \mathrm{im}L^{(1)}_m \to \mathrm{im}L^{(2)}_{m-1}$. Hence, (14) has a solution $v_m \in \mathrm{im}L^{(1)}_m$ for all sufficiently large m and $||v_m||_\omega \le C^{-1}||L^{(2)}_{m-1}f||_{1/\omega} \le$ const, because of the well-known estimate

$$||f - L^{(2)}_{m-1}f||_{1/\omega} \le C(f)m^{-r-\lambda}, \tag{16}$$

(see, e.g. [20, Chap. VI, §2], [14, Conclusion 4.4]).

Estimates (9) and (10) are an immediate consequence of (16), Lemmas 1, 2, 4 and the equation

$$v - v_m = -(D+K)^{-1}\{(K-K_m)v_m + (L^{(2)}_{m-1}f - f)\}.$$

Replacing $L^{(2)}_{m-1}$ by $L^{(2)}_m$ and $L^{(2)}_{\omega,0}$ by $L^{(2)}_\omega$ and repeating the preceding argumentations we prove the assertions of Theorems 1 and 2 for $\chi = 0$. □

Proof of Theorem 3 is standard (see [26], Chap. 14, §4.3) by using known estimates for the Christoffel numbers (see [21, p. 673]).

THE AIRFOIL EQUATION

The results of the preceding section apply to Prandtl's airfoil equation of the form

$$-\frac{1}{\pi}\int_{-1}^{1} \frac{u'(t)}{t-x}\,dt + \frac{1}{\pi}\int_{-1}^{1} T(x,t)u(t)\,dt = f(x), \tag{17}$$

$-1 < x < 1$, where the unknown function $u \in C[-1,1]$ satisfies the conditions $u' \in L^2_{1/\omega}$ with $\omega(x) = (1-x^2)^{-1/2}$ and

$$u(-1) = u(1) = 0. \tag{18}$$

Obviously, the problem (17), (18) can be rewritten as

$$-\frac{1}{\pi}\int_{-1}^{1} \frac{u'(t)}{t-x}\,dt + \frac{1}{\pi}\int_{-1}^{1} k(x,t)u'(t)\,dt = f(x), \int_{-1}^{1} u'(t)\,dt = 0,$$

where $k(x,t) = \int_t^1 T(x,y)dy$. Thus, we have $\alpha = \beta = -1/2$ and $\chi = 1$. Let us seek an approximate solution in the form

$$u_n(t) = \sqrt{1-t^2}\sum_{k=1}^{n} \zeta_k \frac{U_n(t)}{(t-t_{n,k})U_n'(t_{n,k})}\,.$$

Here $U_n(t)$ denotes the Chebyschev polynomial of the second kind, $U_n(t) = \omega(t)\sin[(n+1)\arccos t]$, and $t_{n,k} = \cos y_{n,k}$ $(y_{n,k} = k\pi/(n+1))$, $k = 1,...,n$, are the roots of $U_n(t)$. With

$$u_n'(t) = -\frac{2}{n+1}\sum_{k=1}^{n} \zeta_k \frac{\sin y_{n,k}}{\sin y}\sum_{m=1}^{n} m \sin m y_{n,k} \cos my,$$

$t = \cos y$, $y \in [0,\pi]$, and the well–known relation (see, e.g., [27])

$$\frac{1}{\pi}\int_{-1}^{1} \frac{\omega(t)T_n(t)}{t-x}\,dt = U_{n-1}(x), n = 0,1,...,$$

where $T_n(t) = \cos[n(\arccos t)]$ is the Chebyschev polynomial of the first kind and $U_{-1} \equiv 0$, we obtain that

$$-\frac{1}{\pi}\int_{-1}^{1} \frac{u_n'(t)}{t-t_{n,k}}\,dt = \sum_{j=1}^{n} a_{kj}\zeta_j\,.$$

Here

$$a_{kj} = \begin{cases} (n+1)/2 & \text{if } k = j, \\ 0 & \text{if } |k-j| = 2,4,6,..., \\ \frac{\sin y_{n,j}}{2(n+1)\sin y_{n,k}}\left(\frac{1}{\tan^2\frac{y_{n,j}+y_{n,k}}{2}} - \frac{1}{\tan^2\frac{y_{n,j}-y_{n,k}}{2}}\right) & \text{if } |k-j| = 1,3,... \end{cases}.$$

Thus, to solve (17), (18) approximately, we arrive at the equations

$$\sum_{j=1}^{n}(a_{kj}+b_{kj})\zeta_j = f(t_{n,k}),\ k=1,...,n, \tag{19}$$

where $b_{kj} = \frac{\sin^2 y_{n,j}}{n+1} T(t_{n,k}, t_{n,j})$.

Theorem 4. *Suppose the homogeneous problem corresponding to (17), (18) has only the trivial solution $u \equiv 0$. Then (19) is uniquely solvable for all sufficiently large n and the following error estimates hold.*

(i) If $k \in C^{r+\lambda}([-1,1]^2)$ then

$$||u_n - u||_\infty \leq c||u_n' - u'||_{1/\omega} = O(n^{-r-\lambda})$$

(ii) If k satisfies (1') with $h \in C^{r+\lambda}([-1,1]^2)$ then

$$\begin{aligned} ||u_n' - u'||_{1/\omega} &= O(n^{-r-\lambda}\log n), \\ \max_{-1\leq t\leq 1} |[u_n'(t) - u'(t)]\sqrt{1-t^2}| &= O(n^{-r-\lambda+\frac{1}{2}}\log n), \end{aligned}$$

the latter subject to the condition $r+\lambda > 1/2$.

This theorem follows from Theorems 1 through 3. Numerical performances of the quadrature method under consideration are presented in [28].

References

[1] Belotserkovski, S.M., Lifanov, I.K., Chislennye metody dlja singuljarnych integral'nych uravnenij, Nauka, Moscow 1985. (Russian)

[2] Criscuolo G., Mastroianni, G., Convergence of Gauss type product formulas for the evaluation of two–dimensional Cauchy principal value integrals, BIT **27** (1987), 72–84.

[3] Criscuolo, G., Mastroianni, G., Convergence of Gauss–Christoffel formula with preassigned node for Cauchy Principal–Value integrals, J. Approx. Theory **50** (1987), 326–340.

[4] Criscuolo G., Mastroianni, G., On the uniform convergence of Gaussian quadrature rules for Cauchy principal value integrals, Numer. Math. **54** (1989), 445–461.

[5] Elliott, D., Orthogonal polynomials associated with singular integral equations having a Cauchy kernel, SIAM J. Math. Anal. **13** (1982), 1041–1052.

[6] Elliott, D., The classical collocation method for singular integral equations, SIAM J. Numer. Anal. **19** (1982), 816–832.

[7] Elliott, D., A comprehensive approach to the approximate solution of singular integral equations over the arc (-1,1), J. Integral Equations and Applications **2**, N^0 1 (1989), 59–94.

[8] Elliott, D., Projection methods for singular integral equations, J. Integral Equations and Applications **2**, N^0 1 (1989), 95–106.

[9] Erdogan, F., Gupta, G.D., Cook, T.S., Numerical solution of singular integral equations, Mech. of Fracture **1** (1973), 368–425.

[10] Freud, G., Orthogonal Polynomials, Pergamon, Elmsford, N.Y. 1971.

[11] Gohberg, I., Krupnik, N., Einführung in die Theorie der eindimensionalen singulären Integraloperatoren, Birkhäuser–Verlag, Basel, Boston, Stuttgart 1979.

[12] Golberg, M.A., The numerical solution of Cauchy singular integral equations with constant coefficients, J. Integral Equations **9** (1985), 127–151.

[13] Junghanns, P., Silbermann, B., Zur Theorie der Näherungsverfahren für singuläre Integralgleichungen auf Intervallen, Math. Nachr. **103** (1981), 199–244.

[14] Junghanns, P., Silbermann, B., The numerical treatment of singular integral equations by means of polynomial approximations, Preprint P–Math–35/86, AdW der DDR, Karl–Weierstraß–Institut für Mathematik, Berlin 1986.

[15] Junghanns, P., Silbermann, B., Numerical analysis of the quadrature method for solving linear and nonlinear singular integral equations, Preprint, Technical University Chemnitz, 1988.

[16] Kalandiya, A.I., Mathematical Methods of Two–dimensional elasticity, Moscow 1985.

[17] Mastroianni, G., Prössdorf, S., Some good nodes for Lagrange interpolation, manuscript.

[18] Mikhlin, S.G., Prössdorf, S., Singular Integral Operators, Akademie–Verlag, Berlin 1986, and Springer–Verlag, Berlin, Heidelberg, New York, Tokyo 1986.

[19] Muskhelishvili, N.I., Singular Integral Equations, P. Noordhoff, Groningen 1953.

[20] Natanson, I.P., Konstruktive Funktionentheorie, Akademie–Verlag, Berlin 1955.

[21] Nevai, P., Mean convergence of Lagrange interpolation III, Trans. Amer. Math. Soc. **282** (1984), 669–698.

[22] Prössdorf, S., Silbermann, B., Projektionsverfahren und die näherungweise Lösung singulärer Gleichungen, Teubner Verlagsges., Leipzig 1977.

[23] Prössdorf, S., Silbermann, B., Numerical Analysis for Integral and Related Operator Equations, Akademie–Verlag, Berlin 1991, and Birkhäuser–Verlag, Basel – Boston – Stuttgart 1991.

[24] Venturino, E., Recent developments in the numerical solution of singular integral equations, J. Math. Anal. Appl. **115** (1986), 239–277.

[25] Mastroianni, G., Prössdorf, S., A quadrature method for Cauchy integral equations with weakly singular perturbation kernel, J. Integral Equations and Applications (submitted).

[26] Kantorowitsch, L.W., Akilow, G.P., Funktionalanalysis in normierten Räumen, Akademie–Verlag, Berlin 1964.

[27] Tricomi, F.G., Integral Equations, Interscience Publishers, New York 1957.

[28] Prössdorf, S., Tordella, D., On an extension of Prandtl's lifting line theory to curved wings, Impact of Computing in Science and Engineering (submitted).

Convergence of Spline Approximation Methods for Periodic Elliptic Pseudodifferential Equations

S. Prössdorf, Karl Weierstraß Institut für Mathematik,

Mohrenstraße 39 , D-O 1086 Berlin , Germany

R. Schneider, FB Mathematik TH- Darmstadt,

Schloßgartenstraße 7, D-W 6100 Darmstadt , Germany

Abstract

We investigate several numerical methods for solving the pseudodifferential equation $Au = f$ on the n-dimensional torus T^n. We examine collocation methods as well as Galerkin-Petrov methods using various periodical spline functions. The considered spline spaces are subordinated to a uniform rectangular or triangular grid. For given approximation method and invertible pseudodifferential operator A we compute a numerical symbol α^C, resp. α^G, depending on A and on the approximation method. It turns out that the stability of the numerical method is equivalent to the ellipticity of the corresponding numerical symbol. The case of variable symbols is tackled by a local principle. Optimal error estimates are established.

1. Introduction

Spline approximation via Galerkin procedure or collocation methods is a widely used numerical technique for solving the boundary integral equations arising from exterior or interior boundary value problems of elasticity, aerodynamics, fluid mechanics, electromagnetism, acoustics, and other engineering applications with the boundary element technique, see for example the surveys [4, 25, 13] and references given there. For domains with smooth boundary the corresponding boundary integral equations turn out to be pseudodifferential equations.

For one-dimensional pseudodifferential equations on closed curves, the error analysis of spline approximation methods is rather complete (see e.g. [15, 25, 21]). In particular, conditions are known to be necessary and sufficient for the stabilty of the corresponding numerical procedure in a scale of Sobolev spaces H^s. Generally, these conditions depend on the order of the equation and the degree of the splines. Moreover, optimal order error estimates are established.

In the higher-dimensional case, convergence of spline approximation methods had been shown only in special cases - the most important beeing the case of Fredholm integral equations of second kind with smooth or weakly singular kernels or the boundary integral equation for Laplace's equation over polyhedral domains (see e.g. [9, 8, 22]). We further mention a paper of Martensen [11] where the trigonometric collocation method for a Fredholm integral equation on the torus has been considered. Despite their prevalance, however, no general approach to the error analysis of spline collocation methods for the numerical solution of multi-dimensional pseudodifferential equations was known until recently. To these equations , the standard results of numerical analysis for Fredholm integral equations cannot be applied, since elliptic pseudodifferential operators are not compact on the corresponding Sobolev spaces.

Quite recently two general techniques of the analysis of collocation by odd degree splines at the nodal points have been introduced, the first on the case of periodic pseudodifferential equations, the second in the case of pseudodifferential equations of order zero in bounded Lipschitz domains. The first approach is due to Arnold and Wendland [3] in the one-dimensional case and has been generalized by Hsiao and Prössdorf [10] to multidimensional equations. It is based on equivalence of the collocation method with a mesh dependent Galerkin method and is quite general, yielding optimal asymptotic rates of convergence in the whole scale of Sobolev spaces H^s for which they hold. The second method goes back to Prössdorf and Rathsfeld [16] in the case of one-dimensional singular integral equations with piecewise continuous coefficients and to Prössdorf and Schneider [24, 20] for zero order pseudodifferential equations over Lipschitz domains in $\mathbb{R}^n$. A crucial point of this method is the observation that the collocation matrices for convolution operators with positively homogeneous symbol coincide with the matrices of the finite section for an infinite Toeplitz matrix, which is generated by a function depending on the symbol of the convolution operator and the degrees of the splines. In [7] this approach is used to analyse the convergence of a piecewise bilinear collocation method for a screen problem. The heart of the convergence analysis for operators with variable symbols is the localization principle for spline approximations of Prössdorf [14] which enables us to deduce the stability of the spline approximation method for the general pseudodifferential operator A from the stability for a family of convolution operators derived from A by freezing its principal symbol.

In this paper we present an analysis of spline collocation and Galerkin - Petrov methods for multidimensional periodic pseudodifferential equations which treat the odd and even degree cases together by exploiting the circulant structure of the matrices corresponding to the discretized equations in the case of homogeneous operators (an approach which goes back to Prössdorf and Schmidt [18]) for the piecewise linear nodal collocation for singular integral equations with Cauchy kernel on the unit circle; to more general pseudodifferential equations on closed curves and different approximation methods, this approach has been generalized by Schmidt [23] and Prössdorf and Rathsfeld [17]. The general case of equations with variable symbols can bee reduced to the case of homogeneous symbol again by using the localization principle of [14].

2. Periodic Pseudodifferential Operators

We first collect some basic facts about periodic pseudodifferential operators following [12]. The set $\mathbb{Z}^n$ forms an additive subgroup of $\mathbb{R}^n$. In the sequel we consider the n-dimensional torus as the quotient group

$$\mathcal{T}^n = \mathbb{R}^n / \mathbb{Z}^n .$$

The torus $\mathcal{T}^n$ can be regarded as an n-dimensional manifold.

A function f is called periodic if

$$f(x+k) = f(x) \quad \text{holds for all} \quad x \in \mathbb{R}^n \ , \ k \in \mathbb{Z}^n .$$

Each periodic function induces a unique well defined function on $\mathcal{T}^n$. Conversely, every function f defined on $\mathcal{T}^n$ gives rise to a periodic function which is also denoted by f.

Given a function f with compact support in $\mathbb{R}^n$, we define the periodization operator

$$\mathcal{P}f(x) = \sum_{k \in \mathbb{Z}^n} f(x+k) . \tag{2.1}$$

For an open set Ω we write $\Omega \subset\subset \mathbb{R}^n$ if Ω satisfies

$$\overline{\Omega} \cap (k + \overline{\Omega}) = \emptyset \quad \text{for} \quad 0 \neq k \in \mathbb{Z}^n .$$

Define the integral

$$\int_{\mathcal{T}^n} f(x)dx = \int_{[0,1]^n} f(x)dx .$$

By writing

$$(f, \phi) = \int_{\mathcal{T}^n} f(x)\overline{\phi(x)}dx ,$$

we can identify the space of periodic distributions $\mathcal{D}'(\mathcal{T}^n)$, i.e. the dual space to the space of smooth periodic functions $C^\infty(\mathcal{T}^n)$, with the space of C^∞-densities on $\mathcal{T}^n$.

The Fourier transform on $\mathcal{T}^n$ of a function $f \in C^\infty(\mathcal{T}^n)$ is defined by

$$\hat{f}(\xi) = \mathcal{F}_{\mathcal{T}^n} f(\xi) = \int_{\mathcal{T}^n} e^{-2\pi i \langle \xi, x \rangle} f(x)dx \quad , \quad \xi \in \mathbb{Z}^n , \tag{2.2}$$

where $\langle \xi, x \rangle = \xi^1 x^1 + \cdots + \xi^n x^n$ is the Euclidean inner product. Conversely, due to Fourier's inversion formula f can be recovered by the Fourier series

$$f(x) = \sum_{\xi \in \mathbb{Z}^n} \hat{f}(\xi) e^{2\pi i \langle \xi, x \rangle} \quad , \quad x \in \mathcal{T}^n . \tag{2.3}$$

Conventionally, pseudodifferential operators on smooth manifolds are defined by local representations and by use of a partition of unity.

Definition : *The symbol class $S^m_{cl} = S^m_{cl}(\mathbb{R}^n \times \mathbb{R}^n)$ consists of all functions $\sigma \in C^\infty(\mathbb{R}^n \times \mathbb{R}^n)$ which admits an expansion*

$$\sigma(x, \xi) = \sum_{j=0}^{\infty} \sigma^{m_j}(x, \xi) \quad x, \xi \in \mathbb{R}^n , \tag{2.4}$$

where $\sigma^{m_j}(x,\xi)$ are positively homogeneous function of degree $m_j \in \mathbb{R}$, $m_j \searrow -\infty$ as $j \to \infty$, with respect to the variable ξ for $|\xi| \geq 1$. σ_0 is called the principal symbol.

For a symbol $\sigma \in S^m_{cl}$, define the operator $op\sigma : C_0^\infty(\mathbb{R}^n) \to C^\infty(\mathbb{R}^n)$ by

$$op\sigma\ u(x) = \int_{\mathbb{R}^n}\int_{\mathbb{R}^n} e^{2\pi i\langle\xi,x-y\rangle}\sigma(x,\xi)u(y)dyd\xi \ . \tag{2.5}$$

A pseudodifferential operator of order m on $\mathcal{T}^n$, we write $A \in \Psi^m_{cl}(\mathcal{T}^n)$ is a continuous linear map $A : \mathcal{D}(\mathcal{T}^n) \to \mathcal{D}(\mathcal{T}^n)$ such that, for all functions $\phi,\psi \in C_0^\infty(\mathbb{R}^n)$ with $supp\phi, supp\psi \subset\subset \mathbb{R}^n$, there exist $\sigma_{\phi,\psi} \in S^m_{cl}$ and a smoothing operator K satisfying

$$\phi A(\mathcal{P}\psi u) = op\sigma_{\phi\psi}\psi u + \phi K\psi\ u \quad , \quad u \in C_0^\infty(\mathcal{T}^n)\ .$$

The Fourier transform on the group $\mathcal{T}^n$ gives rise to an alternative definition of pseudodifferential operators.

Definition : *Let $m \in \mathbb{R}$. We denote by $S^m_{cl}(\mathcal{T}^n)$ the restriction of all symbols $\sigma \in S^m_{cl}$ onto $(\mathcal{T}^n \times \mathbb{Z}^n)$. The function $\sigma \in S^m_{cl}(\mathcal{T}^n)$ is called a global symbol on $\mathcal{T}^n$ of order m . For a given global symbol $\sigma \in S^m_{cl}(\mathcal{T}^n)$ we define the operator $\sigma(x,D)$ by*

$$\sigma(x,D)u(x) = \sum_{\xi\in\mathbb{Z}^n} e^{2\pi i\langle\xi,x\rangle}\sigma(x,\xi)\hat{u}(\xi)\ ,\ u \in C^\infty(\mathcal{T}^n)\ ,\ supp u \subset\subset \mathbb{R}^n\ . \tag{2.6}$$

An operator A is called a global pseudodifferential operator on $\mathcal{T}^n$ of order $m \in \mathbb{R}$ if there exist a symbol $\sigma \in S^m_{cl}(\mathcal{T}^n)$ and a smoothing operator K, given by $Ku(x) = \int_{\mathcal{T}^n} k(x,y)u(y)dy$ with $k \in C^\infty(\mathcal{T}^n \times \mathcal{T}^n)$, such that $A = \sigma(x,D) + K$.

The main theorem in [12] asserts that every pseudodifferential operator on $\mathcal{T}^n$ can be described by a global pseudodifferential operator and vice versa. A fact which has been discovered by Agranovich [2, 1] for $n = 1$ and by McLean [12] for $n \geq 1$. Here we need a corresponding version for classical pseudodifferential operators.

THEOREM 2.1 *For every pseudodifferential operator $A \in \Psi^m_{cl}(\mathcal{T}^n)$ there exists $\sigma \in S^m_{cl}(\mathcal{T}^n)$ and a smoothing operator K such that $A = \sigma(x,D) + K$. Conversely, if $\sigma \in S^m_{cl}(\mathcal{T}^n)$ then $\sigma(x,D) \in \Psi^m_{cl}(\mathcal{T}^n)$.*

In [19] we consider a slightly more general class of operators. There we require that at least the principal symbol has to be homogeneous.

As a consequence of Theorem 2.1 the calculus of pseudodifferential operators on $\mathcal{T}^n$ can be carried over to global pseudodifferential operators on $\mathcal{T}^n$.

Let us define the function $\xi \mapsto \langle\xi\rangle$, $\xi \in \mathbb{Z}^n$, by $\langle\xi\rangle = |\xi|$ if $\xi \neq 0$ and $\langle\xi\rangle = 1$ for $\xi = 0$. For $s \in \mathbb{R}$, we define the Sobolev spaces $H^s(\mathcal{T}^n)$

$$H^s(\mathcal{T}^n) = \{u \in \mathcal{D}'(\mathcal{T}^n) : \langle D\rangle^s u \in L^2(\mathcal{T}^n)\}\ . \tag{2.7}$$

A pseudodifferential operator $A \in \Psi^m(\mathcal{T}^n)$ maps

$$A : H^s(\mathcal{T}^n) \to H^{s-m}(\mathcal{T}^n) \quad , \quad s \in \mathbb{R}\ , \tag{2.8}$$

boundedly. In order to apply localization techniques, we use Seeley's lemma [19].

3. Spline Spaces and Approximation Methods for Solving Periodic Pseudodifferential Equations

In our approach several spline spaces of periodic piecewise polynomial functions subordinated to a uniform grid are considered.

On $\mathcal{T}^n$ we introduce a rectangular grid

$$\square^N = \{x_k \in \mathcal{T}^n : x_k = (\frac{k}{N}), k = (k^1, \dots, k^n), 0 \le k^j < N \text{ for } 1 \le j \le n\} .$$

For $n = 2$ we obtain a triangulation Δ^N by dividing each rectangle $[\frac{k^1}{N}, \frac{k^1+1}{N}] \times [\frac{k^2}{N}, \frac{k^2+1}{N}]$ through its diagonal from the point $(\frac{k^1}{N}, \frac{k^2}{N})$ to $(\frac{k^1+1}{N}, \frac{k^2+1}{N})$ into two triangles $\Delta^N_{k,1}$ and $\Delta^N_{k,2}$.

Let $x^j \mapsto \phi_0^{1,0}(x^j)$ be the characteristic function of the interval $(-\frac{1}{2}, \frac{1}{2})$. We define (B-)splines by the d^j-times convolution

$$\phi_0^{1,d^j}(x^j) = \phi_0^{1,0} \star \dots \star \phi_0^{1,0}(x^j) \tag{3.9}$$

and their tensor product

$$\phi_0^{1,d}(x) = \prod_{j=1}^{n} \phi_0^{1,d^j}(x^j) , \tag{3.10}$$

where $d = (d^1, \dots, d^n)$. For $N \in I\!N$ and $k \in Z\!\!Z^n$ we set

$$\phi_k^{N,d}(x) = \phi_0^{1,d}(Nx - k) . \tag{3.11}$$

Define

$$\varphi_k^{N,d} = \mathcal{P}\phi_k^{N,d} , \tag{3.12}$$

provided that

$$\text{supp}\phi_k^{N,d} \subset\subset I\!R^n \tag{3.13}$$

holds. For the set of nodal points

$$\omega^N = \{k \in Z\!\!Z^n : 0 \le \frac{k^j}{N} < 1 , 1 \le j \le n\} , \tag{3.14}$$

the spline space $S^{N,d}(\square^N)$ consists of piecewise multipolynomial functions

$$S^{N,d}(\square^N) = \text{span}\{\varphi_k^{N,d} : k \in \omega^N\} . \tag{3.15}$$

We introduce the Fourier transform on $I\!R^n$ by

$$\mathcal{F}_{I\!R^n} u(\xi) = \int_{I\!R^n} e^{-2\pi i \langle \xi, x \rangle} u(x) dx \quad , \quad u \in C_0^\infty(I\!R^n) . \tag{3.16}$$

Let $\varphi_k^{N,d}$ satisfy (3.13). Then its Fourier coefficients have the form

$$\widehat{\varphi_k^{N,d}}(\xi) = N^{-n} e^{-2\pi i \langle \frac{k}{N}, \xi \rangle} \prod_{j=1}^{n} [\frac{N \sin \frac{\pi \xi^j}{N}}{\pi \xi^j}]_\star^{d^j+1} \quad , \quad \xi \in Z\!\!Z^n , \tag{3.17}$$

where we used

$$[\frac{N\sin\frac{\pi\xi^j}{N}}{\pi\xi^j}]_\star = \begin{cases} \frac{N\sin\frac{\pi\xi^j}{N}}{\pi\xi^j} & \text{if } \xi^j \neq 0 \\ 1 & \text{if } \xi^j = 0 \,. \end{cases} \tag{3.18}$$

For $n = 2$ we define

$$\phi_0^1(x^1,x^2) = \int_{-\frac{1}{2}}^{\frac{1}{2}} \phi_0^{1,0}(x^1-t,x^2-t)dt \tag{3.19}$$

and

$$\phi_k^N(x) = \phi_0^1(Nx-k) \quad , \quad k \in \omega^N \; , \; N \in I\!N \,. \tag{3.20}$$

If supp $\phi_k^N \subset\subset I\!R^n$ holds we apply the periodization operator to define

$$\varphi_k^N = \mathcal{P}\phi_k^N \tag{3.21}$$

and introduce the spline spaces

$$S^N(\Delta^N) = \{\varphi_k^N : k \in \omega^N\} \,. \tag{3.22}$$

The finite dimensional spaces $X^N = \text{span}\{x_k^N : k \in \omega^N\}$ considered in the present paper are either $S^{N,d}(\square^N)$ or $S^N(\Delta^N)$ with $x_k^N = \varphi_k^{N,d}$, resp. $x_k^N = \varphi_k^N$. In the sequel we write $\underline{d} = \min\{d^j : j = 1,\ldots,n\}$ if $X^N = S^{N,d}(\square^N)$ and $\underline{d} = 1$ if $X^N = S^N(\Delta^N)$ (provided $n = 2$).

Consider now the pseudodifferential equation on $\mathcal{T}^n$ of the form

$$Au = f \quad , \tag{3.23}$$

where $A \in \psi^\mu{}_h(\mathcal{T}^n)$, $\text{Re}\,\mu = m$, and $f \in H^{s-m}(\mathcal{T}^n)$ are given and $u \in H^s(\mathcal{T}^n)$ is the unknown solution. The collocation scheme for approximately solving equation (3.23) with continuous f is defined as follows. Determine the trial function $u^N = \sum_{k'\in\omega^N} u_{k'}x_{k'}^N \in X^N$, $X^N \subset H^s(\mathcal{T}^n)$, by solving the system of linear equations

$$Au^N(x_k) = f(x_k) \quad , \quad x_k = (\frac{k}{N}) \quad , \quad \text{for all } \; k \in \omega^N \tag{3.24}$$

for the unknown coefficients $u_{k'}$. Note that if $d = (d^1,\ldots,d^1)$ with odd d^1 then the collocation points are just the grid points of $\square^N$ (nodal-point collocation). But for even d^1 the collocation points are the centers of the cells (center- or mid-point collocation). To define the more general ϵ- collocation method let $\epsilon = (\epsilon^1,\ldots,\epsilon^n)$ with $0 \leq \epsilon^j < 1$, $1 \leq j \leq n$. Then we introduce the translation operator τ_ϵ defined by $\tau_\epsilon u(x) = u(x-\epsilon)$, $u \in \mathcal{D}(\mathcal{T}^n)$. The ϵ−collocation is given by the linear equations

$$Au^N(x_k + \frac{\epsilon}{N}) = f(x_k + \frac{\epsilon}{N}) \quad , \quad x_k = (\frac{k}{N}) \quad , \quad \text{for all } \; k \in \omega^N \,. \tag{3.25}$$

We remark that this definition agrees with the definition given by Schmidt [23] only if d^j is odd. However, ϵ^j differs from that defined in [23] by $\frac{1}{2}$ for even d^j.

If the finite dimensional space

$$Y_\epsilon^N = \text{span}\{y_k^N = \tau_{\frac{\epsilon}{N}} x_k^N : k \in \omega^N\} \,,$$

where $\{x_k^N\}_{k\in\omega^N}$ is a basis in a suitable spline space Y^N, is used as the space of test functions, the Galerkin-Petrov scheme is defined by

$$(Au^N, v^N) = (f, v^N) \quad \text{for all} \quad v^N \in Y_\epsilon^N \tag{3.26}$$

provided (3.26) is well defined.

All these methods have in common that they can be interpreted as projection methods. By $P_{X^N}^{s,N} : H^s(\mathcal{T}^n) \to X^N$ we denote the orthogonal projection in $H^s(\mathcal{T}^n)$. The interpolation projection onto Y_ϵ^N will be denoted by $\Pi_{Y_\epsilon^N}^N$. In the sequel we sometimes write $P_\epsilon^{s,N}$ instead of $P_{Y_\epsilon^N}^{s,N}$ (resp. Π_ϵ^N for $\Pi_{Y_\epsilon^N}^N$). Further, if $\epsilon = 0$ the indication of ϵ will be suppressed.

The crucial property in the theory of projection methods is the stability. For the collocation method, stability means that the estimate

$$\|\Pi_{Y_\epsilon^N}^N Au^N\|_{H^{s-m}(\mathcal{T}^n)} \geq C\|u^N\|_{H^s(\mathcal{T}^n)} \quad , \quad \text{for all} \quad u^N \in X^N \ , \tag{3.27}$$

holds, where Y_ϵ^N is a suitable subspace and C is a positive constant which does not depend on $N \in I\!N$.

4. The Main Results

Let $X^N = S^{N,d}(\square^N)$, resp. $X^N = S^N(\Delta^N)$ and $Y_\epsilon^N = \tau_{\frac{\epsilon}{N}} S^{d',N}(\square^N)$, or $Y_\epsilon^N = \tau_{\frac{\epsilon}{N}} S^N(\Delta^N)$. To lessen the notational burden we write $P_\epsilon^{s,N}$ instead of $P_{Y_\epsilon^N}^{s,N}$, and Π_ϵ^N for $\Pi_{Y_\epsilon^N}^N$.

Further we make the following assumptions which are supposed in each of the following theorems.

Assumption : *Let $d = (\underline{d}, \ldots, \underline{d})$, $X^N = S^{N,d}(\square^N)$, resp. $X^N = S^N(\Delta^N)$ ($n = 2$), and $Y_\epsilon^N = \tau_{\frac{\epsilon}{N}} S^{d',N}(\square^N)$, or $Y_\epsilon^N = \tau_{\frac{\epsilon}{N}} S^N(\Delta^N)$. Further we suppose that in the case of the collocation scheme*

$$s < \underline{d} + \frac{1}{2} \quad , \quad 0 \leq s - m < \underline{d'} + \frac{1}{2} \quad , \quad \underline{d} \geq \frac{n}{2} \tag{4.28}$$

holds. For the Galerkin Petrov scheme we assume

$$m - \underline{d'} - 1 \leq s < \min\{\underline{d}, \underline{d'}\} + \frac{1}{2} \ . \tag{4.29}$$

The last inequality in (4.28) ensures that there exists an $s \in I\!R$ such that the corresponding interpolation projections are uniformly bounded in $H^s(\mathcal{T}^n)$. Note that the choice of the spline degree d' does not effect the collocation equations (3.25). Therefore in this case $\underline{d'}$ can be choosen sufficiently large.

Definition : *The periodic functions α^C and α^G defined by Table 1 are called the numerical symbols of the approximation methods (3.25) and (3.26), respectively. The numerical symbol α^C (resp. α^G) is called* **elliptic** *if there exists a constant $C > 0$ such that*

$$|\alpha^C(y, \eta)| \geq C|\eta|^m \tag{4.30}$$

Table 1

Method	X^N	Y^N	Numerical Symbol $\alpha(x,\eta)$, $x,\eta \in \mathcal{T}^n\backslash\{0\}$
Collo- cation	$S^{N,d}(\square^N)$		$e^{2\pi i\langle\epsilon,\eta\rangle}\sum^*_{\zeta\in\mathbb{Z}^n}\sigma_0(x,\zeta+\eta)e^{2\pi i\langle\epsilon,\zeta\rangle}$ $\prod_{j=1}^n(-1)^{\zeta^j(d^j+1)}[\frac{\sin\pi\eta^j}{\pi(\zeta^j+\eta^j)}]_\star^{d^j+1}$
method	$S^N(\triangle^N)$		$e^{2\pi i\langle\epsilon,\eta\rangle}\sum_{\zeta\in\mathbb{Z}^n}\sigma_0(x,\zeta+\eta)e^{2\pi i\langle\epsilon,\zeta\rangle}$ $[\frac{\sin\pi\eta^1}{\pi(\zeta^1+\eta^1)}]_\star[\frac{\sin\pi\eta^2}{\pi(\zeta^2+\eta^2)}]_\star$ $[\frac{\sin\pi(\eta^1+\eta^2)}{\pi(\zeta^1+\zeta^2+\eta^1+\eta^2)}]_\star$
	$S^{N,d}(\square^N)$	$\tau_{\frac{\epsilon}{N}}S^{N,d'}(\square^N)$	$e^{2\pi i\langle\epsilon,\eta\rangle}\sum_{\zeta\in\mathbb{Z}^n}\sigma_0(x,\zeta+\eta)e^{2\pi i\langle\epsilon,\zeta\rangle}$ $\prod_{j=1}^n(-1)^{\zeta^j(d^j+d'^j)}[\frac{\sin\pi\eta^j}{\pi(\zeta^j+\eta^j)}]_\star^{d^j+1}[\frac{\sin\pi\eta^j}{\pi(\zeta^j+\eta^j)}]_\star^{d'^j+1}$
Petrov -	$S^{N,d}(\square^N)$	$\tau_{\frac{\epsilon}{N}}S^N(\triangle^N)$	$e^{2\pi i\langle\epsilon,\eta\rangle}\sum_{\zeta\in\mathbb{Z}^n}\sigma_0(x,\zeta+\eta)e^{2\pi i\langle\epsilon,\zeta\rangle}$ $[\frac{\sin\pi\eta^1}{\pi(\zeta^1+\eta^1)}]_\star[\frac{\sin\pi\eta^2}{\pi(\zeta^2+\eta^2)}]_\star$ $[\frac{\sin\pi(\eta^1+\eta^2)}{\pi(\zeta^1+\zeta^2+\eta^1+\eta^2)}]_\star$ $\prod_{j=1}^n(-1)^{\zeta^j(d^j+1)}[\frac{\sin\pi\eta^j}{\pi(\zeta^j+\eta^j)}]_\star^{d^j+1}$
Galerkin	$S^N(\triangle^N)$	$\tau_{\frac{\epsilon}{N}}S^{N,d'}(\square^N)$	
method	$S^N(\triangle^N)$.	$\tau_{\frac{\epsilon}{N}}S^N(\triangle^N)$	$e^{2\pi i\langle\epsilon,\eta\rangle}\sum_{\zeta\in\mathbb{Z}^n}\sigma_0(x,\zeta+\eta)e^{2\pi i\langle\epsilon,\zeta\rangle}$ $\lvert[\frac{\sin\pi\eta^1}{\pi(\zeta^1+\eta^1)}]_\star[\frac{\sin\pi\eta^2}{\pi(\zeta^2+\eta^2)}]_\star$ $[\frac{\sin\pi(\eta^1+\eta^2)}{\pi(\zeta^1+\zeta^2+\eta^1+\eta^2)}]_\star\rvert^2$

holds for all $y\in\mathcal{T}^n$, $-\frac{1}{2}<\eta^j<\frac{1}{2}$, $\eta\neq 0$ *and* $j=1,\dots,n$.

Notice that the numerical symbols α^C and α^G depend on the symbol of the pseudodifferential operator A as well as on the employed collocation and Galerkin-Petrov method, respectively.

Proposition 4.2 *If the numerical symbol of the pseudodifferential operator A is elliptic then A is elliptic.*

Remark : In general, the ellipticity of the pseudodifferential operator is not sufficient for the numerical symbol to be elliptic (see e.g. [23]).

We now give some examples of elliptic numerical symbols.

Proposition 4.3 *Consider the rectangular grid $\square^N$. Assume the pseudodifferential operator $\sigma(x,D)\in\Psi^m_{cl}(\mathcal{T}^n)$ is strongly elliptic, i.e.*

$$\mathrm{Re}\,\sigma_0(x,\xi)|\xi|^{-m}\geq C>0\quad,\quad\xi\in\mathbb{Z}^n\,. \tag{4.31}$$

Then the numerical symbol α^C of the collocation method is elliptic, provided $\epsilon=0$ and all d^j, $j=1,\dots,n$, are odd. The numerical symbol α^G of the Galerkin-Petrov method is elliptic if $\epsilon=0$ and $d^j+d'^j$ is even for all $j=1,\dots,n$.

Remark : From a recent result founded by Costabel [5, 6] it follows that in case $\sigma(\xi)=|\xi|^{-1}$, $\xi\neq 0$, and $n=2$ the numerical symbol of the collocation method α^C is elliptic if d^j, $j=1,2$ are even and $\epsilon=0$ (center point collocation).

The next theorems emphazise the importance of the numerical symbol.

THEOREM 4.4 *Suppose the pseudodifferential operator $A \in \Psi^m_{cl}(\mathcal{T}^n)$ is invertible in $\mathcal{L}(H^s(\mathcal{T}^n), H^{s-m}(\mathcal{T}^n))$. Then the collocation scheme (3.25) is stable in $(H^s(\mathcal{T}^n), H^{s-m}(\mathcal{T}^n))$, i.e.*

$$\|\Pi^N_\epsilon A u^N\|_{s-m} \geq C\|u^N\|_s \ , \quad u^N \in X^N \ , N > N_0 \in I\!N \ , \tag{4.32}$$

with $C > 0$, if and only if the numerical symbol α^C is elliptic.

Incorporating the known regularity results for pseudodifferential equations, we obtain quasioptimal convergence.

COROLLARY 4.5 *Let $A : H^s(\mathcal{T}^n) \to H^{s-m}(\mathcal{T}^n)$ be an invertible pseudodifferential operator in $\Psi^m_{cl}(\mathcal{T}^n)$. For $f \in H^{t-m}(\mathcal{T}^n)$, $t-m > \frac{n}{2}$, $m \leq s \leq t$, $t \leq \underline{d}+1$, and $s < \underline{d}+\frac{1}{2}$ the error between the exact solution $u \in H^s(\mathcal{T}^n)$ and the approximate solution $u^N \in X^N$ is bounded by*

$$\|u - u^N\|_s \leq CN^{s-t}\|u\|_t \ . \tag{4.33}$$

Similar results are valid for the Galerkin-Petrov method.

THEOREM 4.6 *Suppose the pseudodifferential operator $A \in \Psi^m_{cl}(\mathcal{T}^n)$, is invertible in $\mathcal{L}(H^s(\mathcal{T}^n), H^{s-m}(\mathcal{T}^n))$. Then the Galerkin-Petrov scheme (3.26) is stable in $(H^s(\mathcal{T}^n), H^{s-m}(\mathcal{T}^n))$, i.e.*

$$\|P^*_{Y^N_\epsilon} A u^N\|_{s-m} \geq C\|u^N\|_s \ , \quad u^N \in X^N \ , N > N_0 \in I\!N \ , \tag{4.34}$$

with $C > 0$, if and only if the corresponding numerical symbol α^G is elliptic.

COROLLARY 4.7 *Let $A : H^s(\mathcal{T}^n) \to H^{s-m}(\mathcal{T}^n)$ be an invertible pseudodifferential operator in $\Psi^m_{cl}(\mathcal{T}^n)$. For $f \in H^{t-m}(\mathcal{T}^n)$, $m - \underline{d}' - \frac{1}{2} < t \leq \underline{d}+1$, $s \leq t$ and $m - \underline{d}' - 1 \leq s < \min\{\underline{d}, \underline{d}'\} + \frac{1}{2}$, the error between the exact solution $u \in H^s(\mathcal{T}^n)$ and the approximate solution $u^N \in X^N$ is bounded by*

$$\|u - u^N\|_s \leq CN^{s-t}\|u\|_t \ . \tag{4.35}$$

5. Outline of the proof of the main theorem

The complete proofs of Theorems 4.4 and 4.6 are worked out in [19]. Here we give only a short outline of the methods which have been used in [19]. For example, the proof of the stability result for the collocation method, namely Theorem 4.4, consists of seveveral steps:

- **Step 1 : Homogeneous operators**

  First we suppose that $A = \sigma(D)$, where the symbol is homogeneous of degree $m \in I\!R$. Here it important to observe that the collocation scheme (3.25) as well as the Galerkin-Petrov scheme (3.26) applied to the equation

  $$\sigma(D)u = f \tag{5.36}$$

  give rise to (higher-dimensional) circulant matrices. These matrices are of the particular form

  $$\mathbf{A} = (a_{kk'}) = (a_{[k-k']}) \ ,$$

where $[k-k'] = k-k'+lN \in \omega^N$ with suitable (and uniquely defined) $l = l(k,k') \in \mathbb{Z}^n$. Such matrices can be diagonalized by

$$\mathbf{VAV}^{-1} = (\alpha_k \delta_{kk'}) \tag{5.37}$$

with the unitary matrices

$$\mathbf{V} = (v_{kk'}) = N^{-\frac{n}{2}}(e^{\frac{2\pi i}{N}\langle k,k'\rangle}) \ . \tag{5.38}$$

Moreover the eigenvalues α_k of $\mathbf{A}$ are given by the formula

$$\alpha_k = \sum_{k' \in \omega^N} a_{k'} e^{-\frac{2\pi i}{N}\langle k,k'\rangle} \ . \tag{5.39}$$

For the collocation method we obtain that the eigenvalues are given by the corresponding numerical symbol α^C.

LEMMA 5.8 *Let $\phi_0 = \phi_0^{1,d}$ be given by (3.12) in the case of $X^N = S^{N,d}(\square^N)$ or $\phi_0 = \phi_0^1$ be given by (3.21) if $X^N = S^N(\Delta^N)$.*

If $\underline{d} > m$ then the eigenvalues α_k^C, $k \in \omega^N$, of the matrices arising from the (ϵ-)collocation scheme (3.25) are $\alpha_0^C = 1$ and

$$\alpha_k^C = N^m \alpha^C(\frac{k}{N}) \quad , \quad k \in \omega^N$$

for $k \neq 0$.

- **Step 2 : Discrete Sobolev norms**

  Next discrete Sobolev norms were introduced. Moreover it was shown in [19] that these norms are uniformly equivalent to the Sobolev norms of the associate spline functions for all $N \in \mathbb{N}$. Further discrete Bessel potential operators have been defined by certain circulant matrices. It turns out that the stability of the employed method is equivalent to the uniform invertibility of a given family of circulant matrices in $l^2(\omega^N)$, $N \in \mathbb{N}$. The eigenvalues of these matrices can be computed explicitly by the numerical symbol α^C together with the symbol of the discrete Bessel potential operators. This gives the desired assertion in Theorem 4.4, provided the pseudodifferential operator is homogeneous, i.e. $A = \sigma(D)$.

- **Step 3 : Local principle**

  The general case can be reduced to the case when A is a homogeneous operator by means of the local principle given in [14]. This local principle can be viewed as a numerical counterpart to the well known procedure of "freezing the coefficients". It requires the validity of several properties of the operators, of the spline spaces and of the projections involved. These properties have been verified in detail in [19, 18]. By means of the local principle one obtains the stability result given in Theorem 4.4.

- **Step 4 : Convergence**

  Finally, the convergence stated in Corollary 4.5 is an immediate consequence of the stability given by Theorem 4.4 together with the well known approximation and inverse properties of the considered spline spaces.

For the Galerkin-Petrov method the analysis is carried out similarly (see [19]). In [19] we further consider some more general circumstances which have been neglected here for reason of compactness.

References

[1] Agranovich, M.: On elliptic pseudodifferential operators on a closed curve. *Trans. Moscow Math. Soc.* **47**, 23 - 74 (1985).

[2] Agranovich, M. S.: Spectral properties of elliptic pseudodifferential operators on a closed curve. *Functional Anal. Appl.* **13**, 279 - 281 (1979).

[3] Arnold, D., Wendland, W.: On asymptotic convergence of collocation methods. *Math. Comp.* **41**, 349 - 381 (1983).

[4] Costabel, M.: Principles of boundary element methods. *Comp. Phys. Reports* **6**, 243–274 (1987).

[5] Costabel, M., McMlean, W.: Spline collocation for strongly elliptic equations on the torus. in preparation.

[6] Costabel, M., Penzel, F., Schneider, R.: A collocation method for screen problems in R^3. erscheint in Tagungsberichte "Elliptic Operators on Singular Manifolds", B.G. Teubner, Leipzig 1991, Breitenbrunn, 30.4. - 4.5. 1990, Manuskript, 8 p.

[7] Costabel, M., Penzel, F., Schneider, R.: Error analysis of a boundary element collocation method for a screen problem in R^3. THD , FB-Math Preprint Nr.1284, 14 p. , Febr., THD Darmstadt 1990.

[8] Golberg, M. e.: *Numerical Solution of Integral Equations.* Plenum Press, New York to appear.

[9] Hackbusch, W.: *Integralgleichungen. Theorie und Numerik.* Teubner, Stuttgart 1989.

[10] Hsiao, G., Prössdorf, S.: A generalization of the Arnold - Wendland lemma to collocation methods for boundary integral equations in $\mathbb{R}^n$. To appear in Math. Meth. Appl. Sci.

[11] Martensen, E.: über eine Methode zum räumlichen Neumannschen Problem mit einer Anwendung für torusartige Berandungen. *Acta Mathematica* **109**, 75 - 135 (1963).

[12] McLean, W.: Local and global description of periodic pseudodifferential operators. manuscript, Sidney 1989, to appear in Math. Nachr.

[13] Nédélec ,J.C. :. Equations intégrales associées aux problèmes aux limites elliptiques dans des domaines de $\mathbb{R}^3$,. In R. Dautray, J.-L. Lions, editors, *Analyse Mathématique et Calcul Numérique pour les Sciences et les Techniques*, chapter XI–XIII. Masson, Paris 1988.

[14] Prössdorf, S.: Ein Lokalisierungsprinzip in der Theorie der Splineapproximationen und einige Anwendungen. *Math. Nachr.* **119**, 239 - 255 (1984).

[15] Prössdorf, S.: Numerische Behandlung singulärer Integralgleichungen. *Z. angew. Math. Mech.* **69**, T 5 - T 13 (1989).

[16] Prössdorf, S., Rathsfeld, A.: A spline collocation method for singular integral equations with piecewise continuous coefficients. *Int. Eqs. and Op. Th.* **7**, 536 - 560 (1984).

[17] Prössdorf, S., Rathsfeld, A.: Stabilitätskriterien für Näherungsverfahren bei singulären Integralgleichungen in l^p. *Z. Anal. Anw.* **6**, 539 - 558 (1987).

[18] Prössdorf, S., Schmidt, G.: A finite element collocation method for singular integral equations. *Math. Nachr.* **100**, 33 - 60 (1981).

[19] Prössdorf, S., Schneider, R.: Spline approximation methods for multidimensional periodical pseudodifferential equations. THD FB-Math Preprint Nr.1341 , November 1990.

[20] Prössdorf, S., Schneider, R.: A spline collocation method for multidimensional strongly elliptic pseudodifferential operators of order zero. THD FB-Math.- Preprint Nr. 1310 , May 1990.

[21] Prössdorf, S., Silbermann, B.: *Numerical Analysis for Integral and Operator Equations.* Birkhäuser Verlag, Basel, Stuttgart 1990. (to appear).

[22] Rathsfeld, A.: The invertibility of the double layer potential operator in the space of continuous functions defined on a polyhedron. Preprint P-MATH-01/90, Karl-Weierstrass-Institut Berlin 1990.

[23] Schmidt, G. Splines und die näherungsweise Lösung von Pseudodifferentialgleichungen auf geschlossenen Kurven. Report R-MATH-09/86, Karl-Weierstrass Institut, Berlin 1986.

[24] Schneider, R.: Stability of a spline collocation method for strongly elliptic multidimensional singular integral equations. THD , FB-Math. Preprint , 21 p. , Dec. to appear Numer. Math. 1272, THD Darmstadt 1989.

[25] Wendland, W. L.: Strongly elliptic boundary integral equations. In: *The State of the Art in Numerical Analysis* (A. Iserles, M. Powell, eds.), pp. 511 - 562. Clarendon Press, Oxford 1987.

CALCULATION OF BLADE–VORTEX INTERACTION OF ROTARY WINGS IN INCOMPRESSIBLE FLOW BY AN UNSTEADY VORTEX–LATTICE METHOD INCLUDING FREE WAKE ANALYSIS

A. Röttgermann, R. Behr, Ch. Schöttl, S. Wagner
Universität der Bundeswehr
Institut für Luftfahrttechnik und Leichtbau
Werner–Heisenberg–Weg 39, D–W–8014 Neubiberg

SUMMARY

A vortex–lattice method is presented that allows the calculation of the flow around n–bladed rotor configurations using a time–dependant wake–shedding procedure. Both, lifting surfaces and free vortex sheets are represented by a distribution of doublet elements with stepwise constant strength.
Beginning with an impulsive start of the configuration the development of all free shear layers is described step by step marching forward in time until a periodic solution for the flowfield and the blade loads is reached.
All boundary conditions — except the kinematic flow condition at the rotor blade collocation points — are implicitely satisfied by the singularity model. A special procedure is used for the update of the geometry of the free vortex sheets so that the numerical problems resulting from the application of Biot–Savart's law can be avoided.
To minimize the computational effort a far wake simplification is introduced that allows coupling of shedded vorticity to large systems of closed vortex rings.
The basic equations, boundary conditions and numerical procedures are discussed. The numerical procedure is validated by comparison with experimental data.

INTRODUCTION

The theoretical approach towards rotary wing aerodynamics is a very interesting task in the field of computational fluid dynamics. Wagner [1] describes all important physical phenomena that are to be considered in theoretical and experimental investigations of helicopter aerodynamics.
The rotor flow is dominated by the trailed and shed vorticity of the rotor blades. Opposite to wings in translatory motion the rotation of the blades causes the maximum of the bound circulation to be located at the outer part of the rotary wing. The steep decrease of the circulation towards zero at the tip produces a strong and concentrated vortex emanating from this region. These tip–vortices and the inboard parts of the wake are deposited in the flow field as the blades rotate and are convected with the local velocity of the fluid. This velocity is a combination of free–stream and induced components.
In general, the geometry of the free vortex sheets consists of distorted interlocking helices, each connected to one blade, skewed downward (see Fig. 1).
Due to the absence of the free–stream velocity in the hovering case the arrangement of the vortical shear layers is very compact and makes this case the most challenging one.

Thereby each blade encounters the vorticity shed from the preceeding blade(s). The variation of the aerodynamic loads caused by blade vortex interaction (BVI) is sensitive to the relative position of the tip vortex and the blade surface (see Fig. 2). According to the complexity of the problem most of the theoretical methods for the description of the rotor flow make use of simplified prescribed wake geometries [3] based on experimental investigations.

The method presented in this paper is a completely analytic tool for the simulation of rotary wing aerodynamics in incompressible and inviscid flow. Up to now the related computer program ROVLM (**RO**tary wings **V**ortex **L**attice **M**ethod) has been successfully applied to helicopter rotors, propellers and horizontal axis wind energy converters [2].

THEORETICAL APPROACH

Governing Equations

Basic assumption of such a model is that the flow is incompressible, inviscid and irrotational — except at the location of the discontinuity surfaces — so that a scalar velocity potential $\Phi = \phi_\infty + \phi$ exists and the continuity equation can be written as Laplace Equation in terms of the pertubation potential ϕ :

$$\nabla^2\phi = \frac{\partial^2\phi}{\partial x^2} + \frac{\partial^2\phi}{\partial y^2} + \frac{\partial^2\phi}{\partial z^2} = 0. \tag{1}$$

A generalisation of Green's theorem allows Eq. (1) to be integrated to give

$$\phi = \underbrace{\frac{1}{4\pi}\int_S \sigma\cdot(\frac{1}{r})\,ds}_{source} + \underbrace{\frac{1}{4\pi}\int_S \mu\vec{n}\cdot\vec{\nabla}(\frac{1}{r})\,ds}_{doublet}. \tag{2}$$

The present method neglects all effects due to thickness of the rotary wings so that only the second part of Equation (2) is used for the representation of the blades and the free vortex sheets.

Applying a low order doublet distribution — i.e. stepwise constant strength μ on each panel $S = S(\vec{r}')$ with unit normal vector $\vec{n}$ — yields

$$\phi(\vec{r},t) = \frac{1}{4\pi}\iint_S \mu(\vec{r}',t)\frac{\vec{n}(\vec{r}',t)\cdot(\vec{r}-\vec{r}')}{|\vec{r}-\vec{r}'|^3}\,dS(\vec{r}',t) \tag{3}$$

for the local pertubation potential at point $P = P(\vec{r})$.

The velocity induced by a set of N quadrilateral panels at a point P can then be obtained using the following expression :

$$\vec{U}(\vec{r},t) = \vec{\nabla}\phi(\vec{r},t) = \frac{1}{4\pi}\sum_{i=1}^{N}\mu_i(\vec{r}_i{}',t)\sum_{j=1}^{4}\frac{\vec{r}_{0j}\times\vec{r}_{1ij}}{(\vec{r}_{0j}\times\vec{r}_{1ij})^2}\cdot\left[\vec{r}_{0j}\left(\frac{\vec{r}_{1ij}}{|\vec{r}_{1ij}|}-\frac{\vec{r}_{2ij}}{|\vec{r}_{2ij}|}\right)\right]. \tag{4}$$

Herein the equivalence of a $\mu = const.$ distribution over a panel and a closed vortex–ring with circulation $\Gamma \equiv \mu$ along the outer edges of the panel is used and Biot–Savart's law is employed.

Numerical Procedure

The rotor blades are discretized into N panels, each carrying a doublet element of constant strength μ (see Fig. 3). Thus, a linear system of N equations can be formulated :

$$\begin{bmatrix} a_{1,1} & a_{1,2} & \cdots & a_{1,N} \\ a_{2,1} & a_{2,2} & \cdots & a_{2,N} \\ a_{3,1} & a_{3,2} & \cdots & a_{3,N} \\ \vdots & \vdots & \ddots & \vdots \\ a_{N,1} & a_{N,2} & \cdots & a_{N,N} \end{bmatrix} \begin{bmatrix} \mu_1 \\ \mu_2 \\ \mu_3 \\ \vdots \\ \mu_N \end{bmatrix} = \begin{bmatrix} \mathcal{R}_1 \\ \mathcal{R}_2 \\ \mathcal{R}_3 \\ \vdots \\ \mathcal{R}_N \end{bmatrix}. \tag{5}$$

The elements of the **A**erodynamic **I**nfluence **C**oefficient (AIC) matrix a_{ki} are calculated according to

$$a_{ki} = \vec{n}_k \cdot \left(\frac{1}{4\pi} \sum_{j=1}^{4} \frac{\vec{r}_{0j} \times \vec{r}_{1ij}}{(\vec{r}_{0j} \times \vec{r}_{1ij})^2} \cdot \left[\vec{r}_{0j} \left(\frac{\vec{r}_{1ij}}{|\vec{r}_{1ij}|} - \frac{\vec{r}_{2ij}}{|\vec{r}_{2ij}|} \right) \right] \right), \tag{6}$$

representing the influence of a quadrilateral panel i with unit singularity strength at collocation point k, see Fig. 4.
The right–hand side vector $\mathcal{R}$ for a panel k is determined by

$$\mathcal{R}_k = -\vec{n}_k \cdot (\vec{U}_\infty(\vec{r}, t) + \vec{\Omega} \times \vec{r}_k) \tag{7}$$

and contains the local velocity due to the sum of rotational motion of the blade and the free–stream velocity perpendicular to the panel surface. That means the kinematic boundary condition — i.e. no flow normal to the panels — is satisfied at the blade collocation points. A first solution for the unknown doublet strengths μ is performed and yields the pertubation potential ϕ_W of the rotor blade, compare Eq. (2).
Along the trailing edge of the blades wake abutting points are introduced, located at the corners of the vortex–rings of the wing–bound lattice, see Fig. 5.
According to Eq. (4) the induced velocities are evaluated at these points by

$$\vec{U}_{l,m}(\vec{r}, t) = \underbrace{\vec{\nabla}\phi_W(\vec{r}, t)}_{rotorblade(s)} + \underbrace{\vec{\nabla}\phi_{VS}(\vec{r}, t)}_{free\ vortex\ sheet(s)} + \underbrace{\vec{U}_\infty(\vec{r}, t).}_{free\ stream\ velocity} \tag{8}$$

Note that the second part of the right hand side of Eq. (8) equals zero — i.e. there is no shedded vorticity — at this moment.
Extending the model into a time–dependant framework a time increment Δt has to be choosen :

$$\Delta t = \frac{\Delta\Psi}{|\vec{\Omega}|} \qquad (\Delta\Psi = 5° \ldots 30°), \tag{9}$$

where $\Delta\Psi$ is the azimuthal angle the blades rotate within each time step. Combining the convection velocities at the wake abutting points from Equation (8) with the time increment Δt a dislocation vector is determined.

$$\vec{r}_{l,m}{}'(t + \Delta t) = \vec{r}_{l,m}(t) + \Delta t \cdot (\vec{U}_{l,m}(\vec{r}, t) + \vec{U}_\infty(\vec{r}, t)). \tag{10}$$

To obtain the final position of the free vortex sheet relatively to the blade the rotational motion of the rotor has to be taken into account.

$$\vec{r}_{l,m}(t + \Delta t) = \vec{r}_{l,m}{}'(t + \Delta t) + \Delta t \cdot (\vec{\Omega} \times \vec{r}_{l,m}{}'(t + \Delta t)). \tag{11}$$

The procedure described above provides a first row of free–wake elements behind the trailing edge, see also Fig. 5. According to Kelvin's Theorem $\partial\Gamma/\partial t = 0$ (dynamic boundary condition), the singularity strength of these wake elements is set equal to the corresponding doublet strength of the preceeding solution at the blade.

$$\mu_{1,m}^{wake}(t+\Delta t) = \mu^{T.E.}(t). \tag{12}$$

Considering the influence of the free vortex sheet(s) $\vec{\nabla}\phi_{VS}$ the right–hand side vector is recalculated extending Eq. (7) to

$$\mathcal{R}_k = -\vec{n}_k \cdot (\underbrace{\vec{U}_\infty(\vec{r},t)}_{free\ stream\ velocity} + \underbrace{\vec{\Omega}\times\vec{r}_k}_{motion\ of\ rotorblade} + \underbrace{\vec{\nabla}\phi_{VS}(\vec{r},t)}_{wake\ induced\ velocities}). \tag{13}$$

The solution of the linear system of equations (5) provides new values for the wing-bound singularity strengths μ. Now the velocities at all corners of the wake elements are recalculated according to Eq. (8). Convection due to the local velocity and the rotational motion yields an additional rank of wake elements behind the blade.
A vortex–ring being once shed at the trailing edge of the blade keeps its circulation throughout the whole computation, i.e. the singularity distribution over the complete wake surface is known at each time.
Repeating this procedure the wake length increases step by step and the bound circulation at the rotory wing changes simultaneously as shown in Fig. 15.
Assuming that there is no geometrical variation of the rotor configuration, the AIC–matrix of the rotor remains constant for all time steps.
The variation of length, shape and position of the free vortex sheets with time effects that the influences of the wake networks onto all examined points — collocation points and points defining the separated vortex sheets — have to be determined again and again. A flow chart of the ROVLM–programm is presented in Fig. 6. Evidently the computational effort increases as the length of the wake grows.

Induced velocities and dislocation of wake networks

In general, during the wake shedding process the inducing vortex–filament and the examined wake–network point or collocation point might probably approach each other very close. In those cases an unrealistic induction results, caused by the $(1/r)$–singularity of Biot–Savart's law.
To avoid irregular distortion of the vortex sheet on the one hand and to guarantee a continous velocity field in the vicinity of the collocation points on the other hand, the following suitable methodologies are introduced.

- To provide realistic amount and direction of the velocity induced by the vortical wake at a *collocation point* the subvortex–technique by B. Maskew [6] is used to prerequisite a proper solution for the bound circulation.

The main idea is to split up a concentrated so–called basic vortex into a system of subvortices distributed on both sides with regard to the local shear layer geometry. The number of subvortices N_{SV} used depends on the normal distance of the point from the vortex sheet:

$$N_{SV} = \text{integer–part–of}\left(1 + \frac{r_v}{r_n}\right), \tag{14}$$

where r_n is the distance between shear layer and point and r_v is the local spacing of the vortex–filaments in the sheet, see Fig. 7. The strength of the subvortices Γ_{SV} is determined by

$$\Gamma_{SV,i} = \Gamma_{\text{basic}}/N_{SV}^2(i-0.5), \tag{15}$$

and the position is obtained by

$$r_{SV,i} = r_{\text{basic}} \pm (N_{SV} + 0.5 - i)\frac{r_v}{N_{SV}}. \quad (16)$$

For the present examinations, the maximal number of subvortices N_{SV} is restricted to 16 vortices on both sides and a local damping radius r_D is used, being 50% of the local subvortex spacing:

$$v_{ind} = v_{ind,BS}\left(1 - \exp\left(-\frac{r^2}{r_D^2}\right)\right), \quad (17)$$

herein $v_{ind,BS}$ is the velocity induced by Biot–Savart's law.

- For the update of the *wake geometry* the induced velocities resulting from Biot–Savart's law are converted into dislocation vectors representing a motion along a circular path around the inducing vortex, compare R. Behr [5].

Therefore, the tangential velocity $\vec{U}_{ind}$ is used to calculate an angular velocity $\omega = |\vec{U}_{ind}|/|\vec{r}|$. The new position of a point is obtained by rotation with the azimuthal angle $\Delta\varphi = \omega \cdot \Delta t$, see Fig. 8.
Discrete vortices being located in direct neighborhood are now enabled to keep their relative position and even the formation of large structures of vortices resulting from accumulation and amalgamation is possible. Especially in hovering cases, where is no free–stream velocity that dominates the transport of the separated vortex sheets, the ability mentioned above is a very important feature that has to be guaranteed for a successful theoretical analysis.

Wake properties and far wake simplification

During the course of different applications of the present method to wind energy converters and helicopter rotors a remarkable phenomenon has been observed. Immediately after the start of the numerical simulation the diameter of the shed wake tube always shows the opposite behaviour as expected for the steady state solution of the problem, i.e. vortical wakes of wind energy converters contract during the first time–steps and the spiral–type vortex sheets behind helicopter rotors show an increasing diameter. When the model approaches steady state conditions a contraction of the wake tube due to the acceleration of the flow is obtained in the case of driven rotors and an expansion is generated by wind–milling rotors.
Of course, this behaviour of the wake tube can be explained by looking at the induced velocities of the locally dominating tip vortices. Fig. 9 points out that the behaviour of the tail end of the wake tube depends only on the sense of rotation of the tip vortices and shows contraction or expansion respectively.
Note, although the steady rotor flow in theory — and in experiment too — is not directly effected by this phenomenon all numerical procedures that are not based on prescribed wake geometries should consider this effect.
During the start–up process of a rotor in hover, as explained in previous sections, this characteristic property of wake interaction yields a dramatic concentration of the vorticity in the vincinity ot the rotor plane. A toroidal structure containing the shed vorticity of a couple of revolutions is generated. When a certain amount of vorticity is gathered within this formation the self-induced downwash forces this vortical torus moving away from the rotor disk. In the present method a lot of wake elements are necessary to form this torus. As the computational effort for treating this region is high, it has been decided to replace this area by two simple vortex–rings, one representing the tip vortices and the other containing the inboard parts of the vortex sheet.

To define the initial locations of these rings a set of radial planes including the rotor axis is introduced. Calculating the intersection points $r_{i,j}$ of all streamwise vortex filaments i with planes j and separately summing up with regard to their sense of rotation of Γ_i according to

$$\vec{r}_j = \frac{\sum_{i=1}^{N_j} \Gamma_i \vec{r}_{i,j}}{\sum_{i=1}^{N_j} \Gamma_i} \tag{18}$$

provide strength and position of the vortex rings substituting the vortical torus. A typical number of planes N_j within the present method is $N_j = 12 \ldots 24$. From now on the rings representing the "far" wake are treated in the same way than all other parts of the free vortex sheets. A typical application of this model is given in the next section.

RESULTS

Fig. 10 shows the time history of the wake development for the start-up procedure of a two–bladed rotor in hover. For the sake of simplicity only the free vortex sheet of one blade is depicted. The azimuthal angle $\Delta\Psi$ is choosen to be 15 degrees per time–step. After approximately one revolution the wake is still concentrated in the close neighborhood of the rotor and even some parts of the vortex sheet are pushed above the rotor disk (Fig. 10a). After 2.5 revolutions ($60 \cdot \Delta t$) the vortex filaments are gathering in a ring vortex due to the starting procedure as described earlier (Fig. 10b). Note that the vortex sheets spread out so that the diameter of the outer contour of the wake is greater than the diameter of the rotor disk (Fig. 10c). Marching forward in time the vortical torus from the start–up procedure is convected downward and the typical helices of vortex sheets with radial contraction and interlocking surfaces are generated by the numerical method (Fig. 10d,e). The geometry of the free vortex sheet after 12 revolutions (288 time–steps) of the rotor is shown in Fig. 10f.
To validate the numerical results an experiment was carried out at the RWTH Aachen [7] visualizing the tip vortex paths during the start–up process of a hovering rotor ($n_b = 2, b = 0.39, r_0 = 0.11, c_i = c_a = 0.054, \alpha_{col} = 10°, \Omega = 26.18$). In Fig. 11 a comparison of theory and experiment is shown. Both figures refer to a situation after 6 revolutions, and clearly demonstrate the outward directed motion of the tail end of the free wake tube.
For a detailed analysis of the flow field after 12 completed revolutions — steady state blade loads are reached — the distribution of vorticity is displayed in a cross section at an azimuthal angle 30 degrees behind the blade, see Fig. 12. The black areas indicate the trace of the tip vortices — maximum values of circulation — the areas in light grey are free of vorticity. The regions plotted in medium grey show the staggered shear layers. Note again the radial contraction of the wake tube close below the rotor disk continuing with a constant diameter and finally ending up in a strong expansion and roll–up of the tail end.
Fig. 13 and 14 concern a 4–bladed rotor with a collective pitch of $\alpha_{col} = 10°$ [8]. The results in Fig. 13 indicate that the maximum circulation value, its radial location, as well as the spanwise distribution of the bound circulation at the trailing edge of the blade are in good agreement with experimental data [8],[9]. In Fig. 14, the radial and axial location of the tip vortex is plotted as a function of azimuth. The position of the tip vortex core in relation to the vortex sheet geometry is obtained according to Eq. (18) using all streamwise discrete vortex lines shed from the trailing edge region between the panel with the maximum value of circulation and the tip of the blade. The determined locations of the tip vortex core define the coordinates of a single tip vortex path compared

to experimental results [8] in Fig. 14. The convergence of the tip vortex geometry towards the steady state conditions of the experiment can clearly be observed. The sharp increase in the axial position after the passage of the following blade is precisely reproduced.
Fig. 15 is related to the rotor used in the experiment of the RWTH Aachen [7] and shows the development of the wing bound circulation in time. The initial increase of the blade doublet distribution is a response to the impulsive start of the rotor analogous to the well-known Wagner function in two-dimensional flow. After about 180 degrees (12 time steps) the blade solutions are for the first time significantly influenced by the wake of the preceeding blade. During the following 2 to 4 revolutions (about 100 time steps) the blade-vortex interaction is dominated by the toroidal structure arising from the starting process. As the vortical torus is convected downward, its impact on the blades deminishes and the loads approach steady state values. The peak seen in Fig. 15 at about 175 time steps is caused by the introduction of the wake simplification model. The disturbances of that substitution vanish during the continuation of the calculation, indicating that the ring vortices, representing the tail end of the wake tube, have the same effect onto the blade loads. The operation mode of the wake simplification procedure is demonstrated in Fig. 16. Assuming a free wake calculation with a dominant vortical torus (Fig. 16a) the tail end of the wake is cut off leaving only about one revolution of the free vortex sheet aft the rotor (Fig. 16b). As previously mentioned, a pair of ring vortices is obtained for each wake from the cut-off region of the vortex sheet (Fig. 16c). The rings have almost the same effect onto the blades and the remaining wake as the original end of the stream tube, but they are consuming only a fraction of the computational time required without the far wake replacement. The two rings are influencing all other points — wake points, collocation points at the blade as well as ring points — and are influenced and distorted by the remaining wakes and blades. The great advantage of this procedure is that the CPU-time formerly needed to compute the self-induced velocities of a large amount of discrete vortex lines is reduced to the investigation of few ring points. In the course of time the tail end of the vortex tube is rolling up outward again and the substitution procedure can be repeated (Fig. 16d).
Fig. 18 gives an impression of the computational time needed during the calculation without far-wake simplification and the CPU-time saved by using the ring vortices. The dashed line indicates the saving of computational effort for an optimized employment of the substitution. The calculations were executed on the Borroughs UNISYS A15 (0.8 MFlops). Installation on a work-station (IBM RISC 6000/530) decreases the computational time by a factor of 7 at least.
Fig. 19 presents the wake geometry of a 4-bladed rotor in forward flight. For clarity only two vortex sheets are shown. As described in detail in Ref. [4], in a rotating coordinate system the relative positions of blade and wake vortices vary periodically, producing a strong variation in the wake induced velocity encountered by the blade and hence in the blade loading, see Fig. 17. The vortex wake of rotary wings in forward flight rolls up in a two-stage process. The vortex filaments of each blade quickly roll up into concentrated lines, as they are shed from the blades. Then the interlocking, overlapping helices in the far wake interact and roll up to form two seperated vortex structures similiar to those of a finite wing in translatory motion. As in this computation there is no cyclic control to balance the loads on the advancing and retreading side in forward flight, the resulting circulations of advancing blades are much stronger and therefore produce a larger fixed wing type tip-vortex system on this side, see Fig. 19.

CONCLUSIONS

A free wake model for n-bladed rotary wing configurations in arbitrary operation conditions has been derived and demonstrated for both hover and forward flight applications. The geometry is restricted to thin rotor blades which may have any shape. Besides the operating conditions the geometrical data defining the rotary wing configuration are the only input data necessary for the theoretical model. The ROVLM code is able to determine the structure of the vortex sheets shed from the rotary wings and to accurately predict the convection of the tip vortices. In addition, the aerodynamic load distribution along the blades is obtained with acceptable accuracy, noting that compressible and viscous effects are excluded by the formulation of the governing equations. The presented far wake simplification allows the application of the method within different stages of rotor design at comparatively low computational costs.
The major task of the ROVLM code is to determine shape and position of free vortex sheets behind rotary wings. However, using this information makes it also possible to calculate the dynamic inflow in the rotor disk and to obtain the induced angles of attack at different radial sections along the rotor blade. Up to now, the ROVLM code has been successfully used within rotor and wind energy converter investigations. Further studies and improvements, concerning the combination with bodies/fuselages and the extension to compressible viscous flow are planned in future work.

REFERENCES

[1] Wagner, S. : *Die Aerodynamik des Hubschraubers — aktuelle Probleme und Lösungsansätze*, Übersichtsvortrag auf Einladung der AG STAB, 7. DGLR–Fach–Symposium über Strömungen mit Ablösung, RWTH Aachen, 7.-9. November 1990, to be published

[2] Schöttl, Ch.; Wagner, S.; Lösch, R.: *Dynamic and Aeroelastic Response of Wind Energy Converters to Aerodynamic Loads*, handbooks, ed. Shock R.; Beurskens J., Brussels B, to be published

[3] Landgrebe, A.J. : *An Analytic and Experimental Investigation of Helicopter Rotor Hover Performance and Wake Geometry Characteristics*, Moffet Field, CA, US-AAMRDL TR-71-24, June 1971

[4] Johnson, W. : *Helicopter Theory*, Princeton University Press, 1980

[5] Behr, R.; Wagner, S. : *A Vortex–Lattice Method for the Calculation of Vortex Sheet Roll–up and Wing–Vortex Interaction*, Finite Approximations in Fluid Mechanics II, Notes on Numerical Fluid Mechanics, Vol. 25, S. 1 – 14, Vieweg–Verlag, Braunschweig 1989

[6] Maskew, B. : *Subvortex Technique for the Close Approach to a Discretized Vortex Sheet*, Journal of Aircraft, Vol. 14, No. 2, pp. 188-193

[7] Müller, R.H.G. : RWTH Aachen, private communication

[8] Favier, D.; Nsi Mba, M.; Barbi, C.; Maresca, C. : *A Free Wake Analysis for Hovering Rotors and Advancing Propellers*, Vertica, Vol. 11, No. 3, pp. 493-511, 1987

[9] Nsi Mba, M.; Meylan, C.; Maresca, C.; Favier, D. : *Radial distribution circulation of a rotor in hover measured by laser velocimeter*, 10th European Rotorcraft Forum, No. 12, Aug. 1984

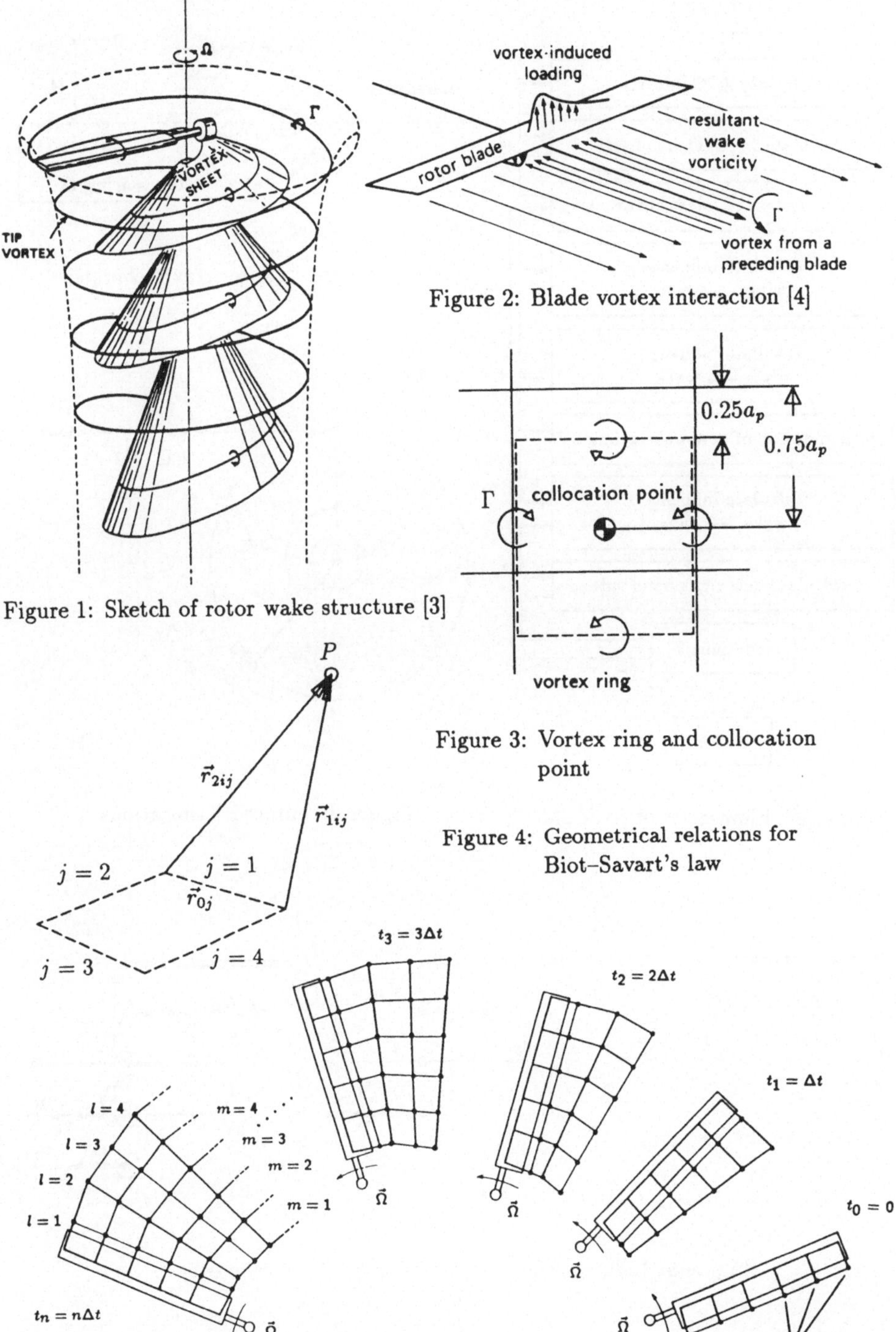

Figure 1: Sketch of rotor wake structure [3]

Figure 2: Blade vortex interaction [4]

Figure 3: Vortex ring and collocation point

Figure 4: Geometrical relations for Biot–Savart's law

Figure 5: Vortex–lattice model of rotor wake

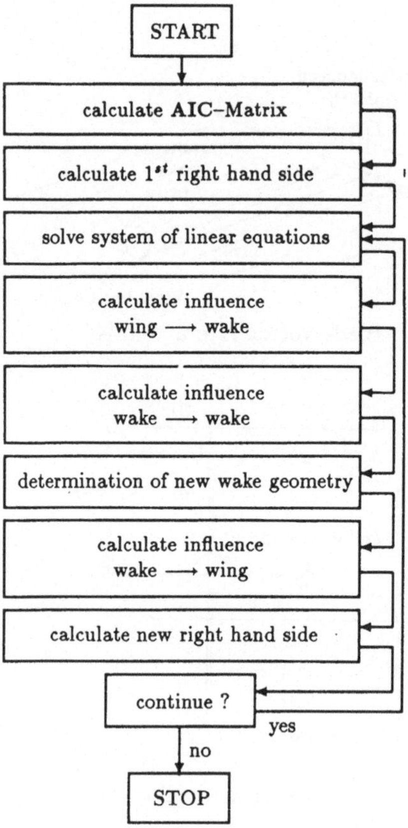

START

calculate AIC–Matrix

calculate 1st right hand side

solve system of linear equations

calculate influence
wing ⟶ wake

calculate influence
wake ⟶ wake

determination of new wake geometry

calculate influence
wake ⟶ wing

calculate new right hand side

continue ?

yes

no

STOP

Figure 6: Flow chart of ROVLM

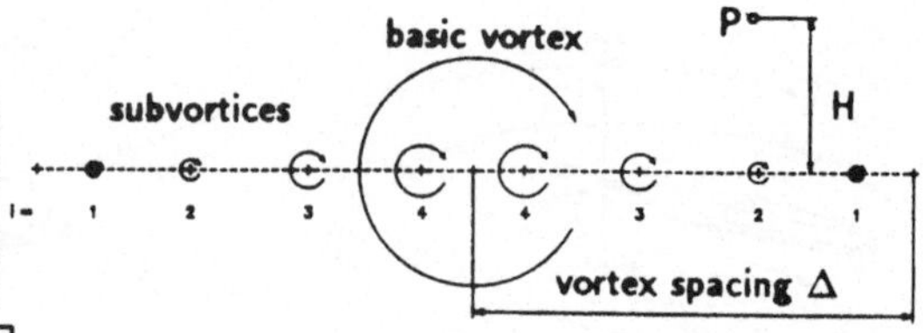

basic vortex

P

H

subvortices

1 2 3 4 4 3 2 1

vortex spacing Δ

Figure 7: Subvortex technique

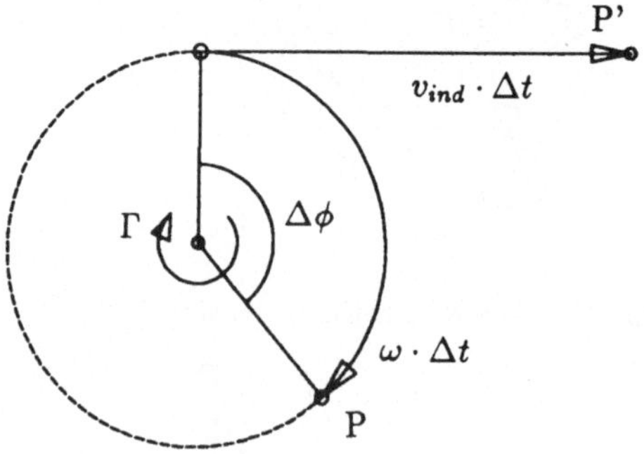

P'

$v_{ind} \cdot \Delta t$

Γ

$\Delta\phi$

$\omega \cdot \Delta t$

P

Figure 8: Induced Dislocations

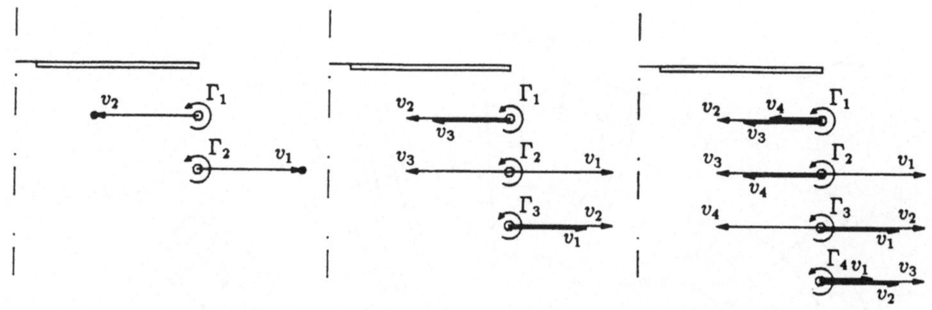

Figure 9: Induced velocities at tail end of the wake tube

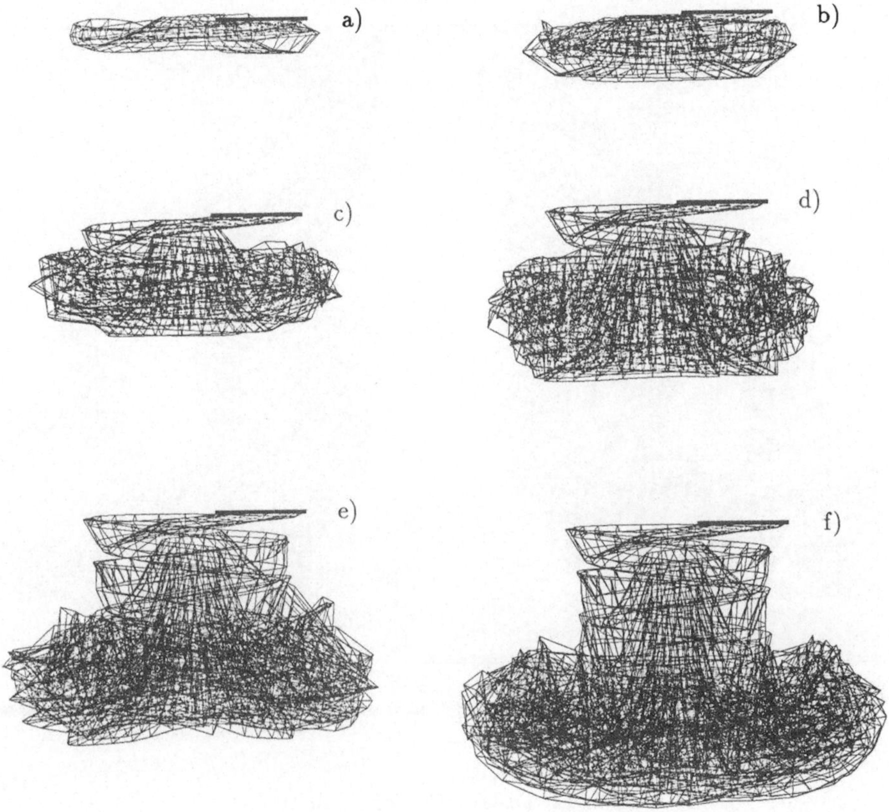

Figure 10: Time history of wake development

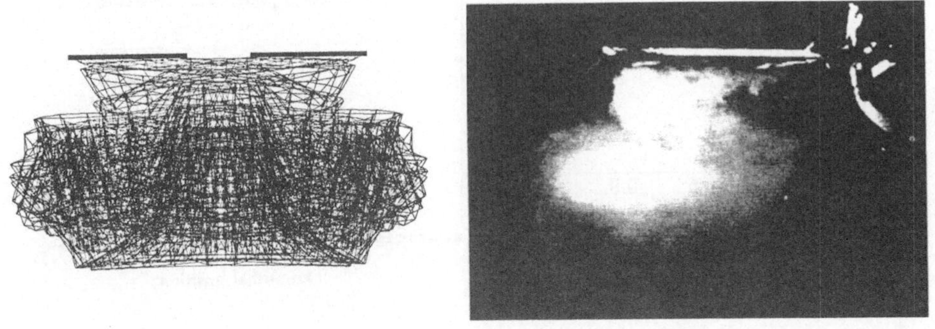

Figure 11: Comparison of theory and experiment [7]

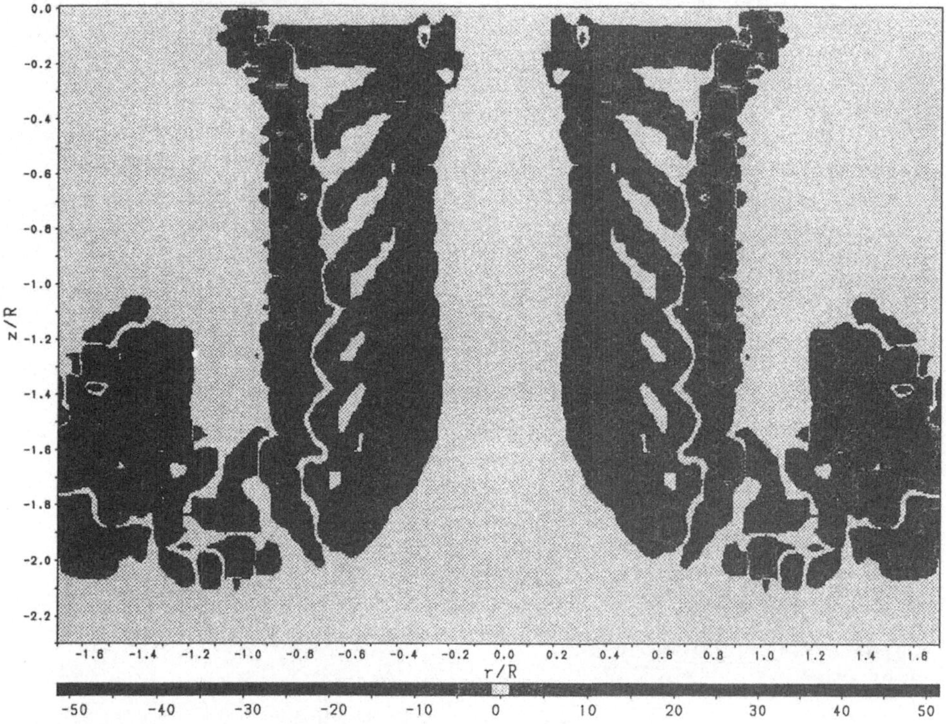

Figure 12: Distribution of vorticity

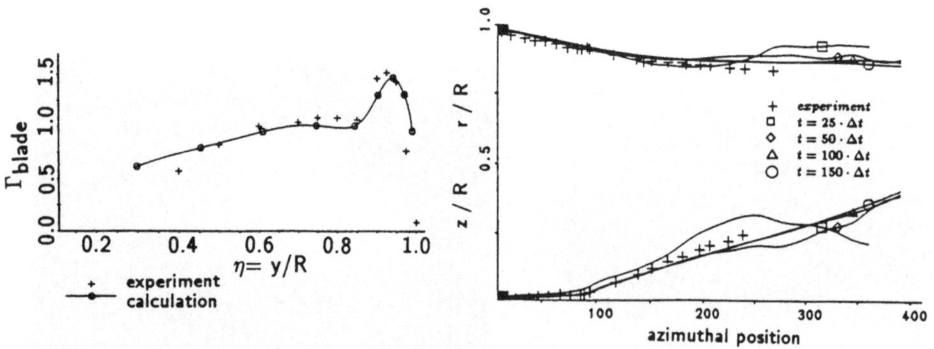

Figure 13: Spanwise circulation distribution

Figure 14: Tip vortex path

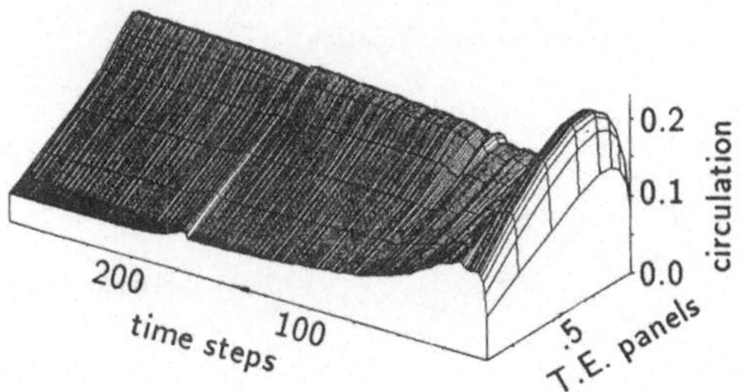

Figure 15: Time history of wing bound circulation in hover

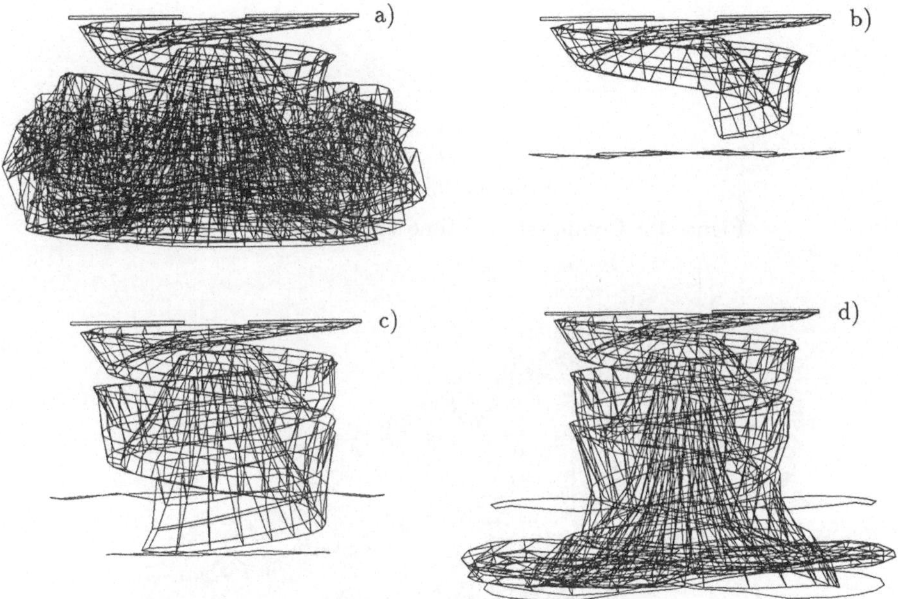

Figure 16: Wake simplification procedure

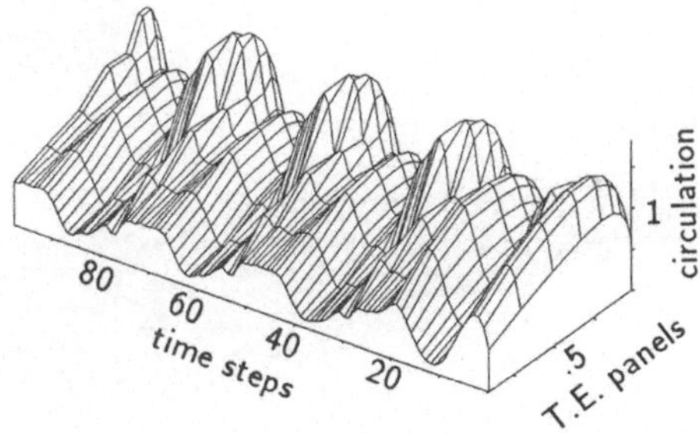

Figure 17: Time history of wing bound circulation in forward flight

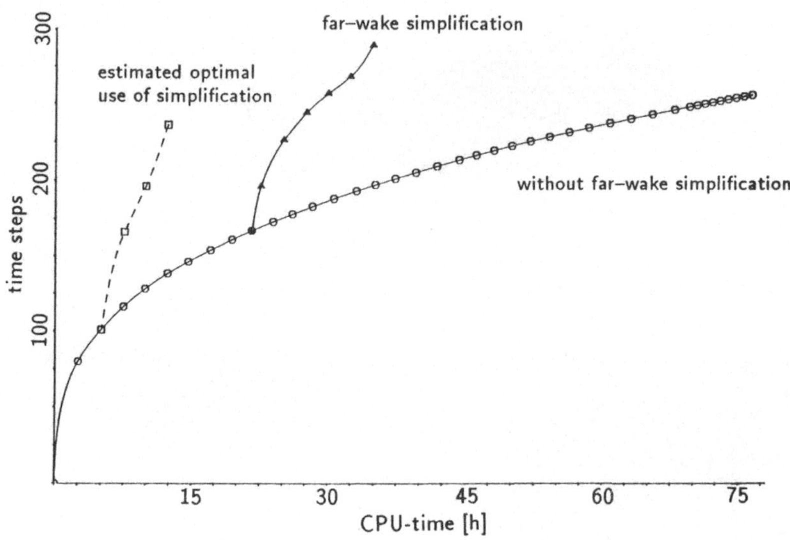

Figure 18: Computational time vs time–steps

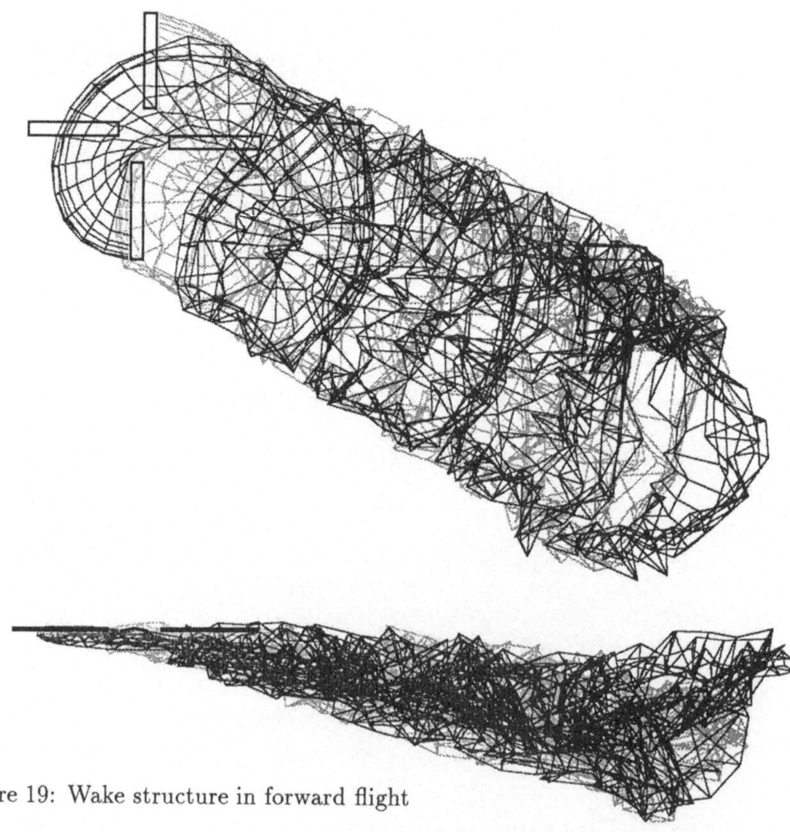

Figure 19: Wake structure in forward flight

NUMERICAL TECHNIQUES FOR THE COUPLING STIFFNESS MATRIX OF FEM AND BEM

E. Schnack
Institute of Solid Mechanics, Karlsruhe University
Kaiserstraße 12, D-7500 Karlsruhe 1, Germany

SUMMARY

Combining the boundary integral equation method (BEM) with the variational formulation (FEM) yields, as a final result, either a symmetric formulation or a positive definite one. In order to develop a coupling procedure which has both of these properties, we consider Courant's idea in elasticity. We convert the strain energy term into a boundary energy. So we have the possibility to define a common total potential energy functional for the BEM and FEM-subregion. This is possible for exterior domain problems, for example in fluid or soil mechanics, but also in solid mechanics, for example for interior domains in elasticity. It leads to a mixed variational formulation with a weak form for the equilibrium and a weak form for the comptability equation. Finally, in the numerical procedure, we have a symmetric, positive definite formulation. For both the static and the kinematic group, we can define the corresponding spaces exactly. This yields some informations about the stability and the convergence of the coupling procedure. Beside the linear elasticity we can follow this idea for elasto-plasticity, too.

Numerical results for two- and three-dimensional test problems give evidence of the advantages of this coupling procedure. The large finite element we have constructed using the BEM yields results of a higher level of accuracy than a classical finite element could do. The following papers give a survey on coupling [1-16], [18-19], [21-23].

CONSTRUCTION OF THE METHOD

We consider the problem of elasticity of a two- or three-dimensional homogeneous, isotropic body with only small deformations. Figure (1) gives the definition of the domain Ω with the two subregions Ω_F and Ω_B. The fundamental equations are the following:

$$\Omega \subset \mathbb{R}^i \quad \text{with } i = 2,3\,. \tag{1}$$

The boundary Γ is defined by:

$$\Gamma := \Gamma_t \cup \Gamma_u. \tag{2}$$

Using the Δ^*-operator, the Navier equation reads:

$$\Delta^*\mathbf{u} := \mu\, \Delta\mathbf{u} + (\lambda + \mu)\ \text{grad div}\ \mathbf{u} = \mathbf{0} \quad \text{in } \Omega. \tag{3}$$

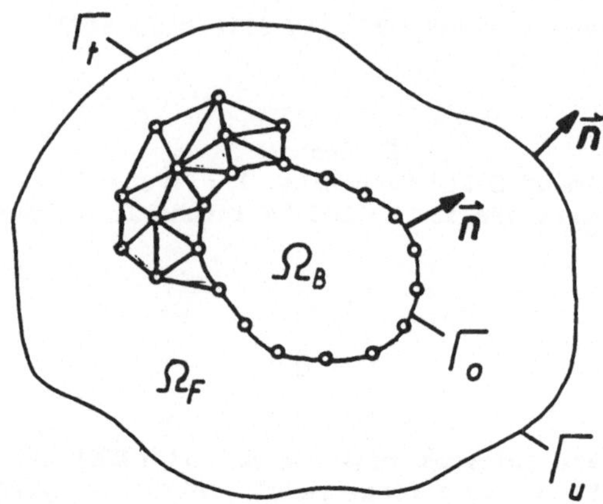

Fig. 1. Embedded macro-element Ω_F: Region of Finite Elements
Ω_B: Region of Boundary Elements

λ and μ are the Lamé-constants. The traction operator T leads to the boundary tractions:

$$T[u]\Big|_{\Gamma_t} := \lambda(\text{div } u)\, n + 2\,\mu \frac{\partial}{\partial n} u + \\ + \mu n \times \text{rot } u\Big|_{\Gamma_t} = \psi \quad \text{on } \Gamma_t, \tag{4}$$

which define the Neumann condition, while the Dirichlet condition is given by:

$$u = \theta \quad \text{on } \Gamma_u. \tag{5}$$

We have the corresponding fundamental problems for the subdomain Ω_F. In that formulation, v^- and v^+ represent the limits of any function v on Γ_0 from Ω_F and Ω_B respectively. The first problem is the following:

$$\Delta^* u_F = 0 \quad \text{in } \Omega_F, \tag{6}$$

$$u_F = \theta \quad \text{on } \Gamma_u, \tag{7}$$

$$T[u_F] = \psi \quad \text{on } \Gamma_t, \tag{8}$$

$$T[u_F^-] = \sigma \quad \text{on } \Gamma_0. \tag{9}$$

For the subdomain Ω_B we have the equations:

$$\Delta^* u_B = 0 \quad \text{in } \Omega_B. \tag{10}$$

$$u_B^+ = \phi \quad \text{on } \Gamma_0. \tag{11}$$

We define the transmission conditions:

$$\sigma = T\left[u_B^+\right] \quad \text{on } \Gamma_0, \tag{12}$$

$$\phi = u_F^- \quad \text{on } \Gamma_0. \tag{13}$$

A central signification is given to the strain energy term U_B of the deformed domain Ω_B. Applying Green's formula, this strain energy term yields:

$$U_B = \frac{1}{2} \int_{\Omega_B} E\left(u_B, u_B\right) \, d\Omega_x = \frac{1}{2} \int_{\Gamma_0} u_B^+ \, T\left[u_B^+\right] \, ds_y. \tag{14}$$

The idea of converting the domain integral into a boundary integral comes originally from Courant. From the computational point of view it has been introduced by Schnack [17]. u_B^+ and $T\left[u_B^+\right]$ are required to constitute a solution satisfying the govnerning Navier equation a priori.

Now we introduce a variational principle by seeking the saddle point of the total potential energy of Ω_F and the boundary energy of Ω_B over a class of admissible functions, see Hsiao [13]. We define the functional Π which represents the total potential energy of Ω:

$$\begin{aligned} \Pi\left(u_F, u_B, \lambda\right) := & \frac{1}{2} \int_{\Omega_F} E\left(u_F, u_F\right) \, d\Omega_x + \frac{1}{2} \int_{\Gamma_0} u_B^+ \, T\left[u_B^+\right] \, ds_y + \\ & + \int_{\Gamma_0} \lambda\left(u_F^- - u_B^+\right) \, ds_y - \int_{\Gamma_t} \psi \, u_F \, ds_y. \end{aligned} \tag{15}$$

We have used the Lagrange multiplier λ to satisfy the continuity requirement for the displacements. Without any loss in general, we assume that Θ equals zero in Equation (5). (v_F, v_B, μ) are the corresponding test functions for $\left(u_F, u_B, \lambda\right)$. The stationary condition $\delta\Pi = 0$ leads to the variational formulation:

Find $\left(u_F, u_B, \lambda\right) \in U_{ad}$ such that:

$$a\left(u_F, v_F\right) + \langle \lambda, v_F^- \rangle = \int_{\Gamma_t} \psi \, v_F \, ds_y, \tag{16}$$

$$\langle \mu, u_F^- - u_B^+ \rangle = 0, \tag{17}$$

$$\langle T\left[u_B^+\right] - \lambda, v_B^+ \rangle = 0, \tag{18}$$

for all: $\left(\mathbf{v}_F, \mathbf{v}_B, \mu\right) \in \mathbf{U}_{ad}$,

where $\langle .,. \rangle$ means the integration over Γ_0.

$a(\mathbf{u}_F, \mathbf{v}_F)$ is defined by:

$$a(\mathbf{u}_F, \mathbf{v}_F) := \int_{\Omega_F} E(\mathbf{u}_F, \mathbf{v}_F)\, d\Omega_x \tag{19}$$

with

$$E(\mathbf{u}_F, \mathbf{v}_F) := (\lambda+\mu - \frac{i-2}{i}\mu) \operatorname{div} \mathbf{u}_F \operatorname{div} \mathbf{v}_F +$$

$$+ \frac{1}{2}\mu \sum_{j\neq k} \left(\frac{\partial v_j^F}{\partial x_k} + \frac{\partial v_j^F}{\partial x_j}\right)\left(\frac{\partial u_j^F}{\partial x_k} + \frac{\partial u_k^F}{\partial x_j}\right) + \frac{1}{i}\mu \sum_{j,k=1}^{i} \left(\frac{\partial v_j^F}{\partial x_k} - \frac{\partial v_k^F}{\partial x_j}\right)\left(\frac{\partial u_j^F}{\partial x_k} - \frac{\partial u_k^F}{\partial x_j}\right),$$

with $i = 2,3$. (20)

$\mathbf{U}_{ad}$ is the appropriate function space for the trial and test functions. With $\lambda = T\left[\mathbf{u}_B^+\right]$ we get the following two equations:

$$a\left(\mathbf{u}_F, \mathbf{v}_F\right) + \langle T\left[\mathbf{u}_B^+\right], \mathbf{v}_F^-\rangle = \int_{\Gamma_t} \psi\, \mathbf{v}_F\, ds_y, \tag{21}$$

$$\langle \mu, \mathbf{u}_F^- \rangle = \langle \mu, \mathbf{u}_B^+ \rangle, \tag{22}$$

with $\boldsymbol{\mu} = T\left[\mathbf{v}_B^+\right]$ we get:

$$\langle T\left[\mathbf{v}_B^+\right], \mathbf{u}_F^- \rangle = \langle T\left[\mathbf{v}_B^+\right], \mathbf{u}_B^+ \rangle. \tag{23}$$

The initial formulation is defined as:

$$\mathbf{u}_B^+ = \mathbf{L}\,\boldsymbol{\beta},\quad T\left[\mathbf{u}_B^+\right] = \mathbf{R}\,\boldsymbol{\beta}, \tag{24}$$

$$\mathbf{v}_B^+ = \mathbf{L}\,\boldsymbol{\alpha},\quad T\left[\mathbf{v}_B^+\right] = \mathbf{R}\,\boldsymbol{\alpha}, \tag{25}$$

with $\boldsymbol{\alpha}$ and $\boldsymbol{\beta}$ being sets of coefficients. $\mathbf{u}_B$ and $\mathbf{v}_B$ have to satisfy the Navier equation (Equation (3)) in Ω.

We have the symmetric property:

$$\langle T\left[\mathbf{u}_B^+\right], \mathbf{v}_B^+ \rangle = \langle T\left[\mathbf{v}_B^+\right], \mathbf{u}_B^+ \rangle. \tag{26}$$

We require for $\mathbf{u}_F^-$ and $\mathbf{v}_F^-$ the same shape functions $\mathbf{N}$:

$$u_F^- = \mathbf{N}\,\boldsymbol{\delta}, \tag{27}$$

$$v_F^- = \mathbf{N}\,\boldsymbol{\gamma}, \tag{28}$$

with $\boldsymbol{\delta}$ and $\boldsymbol{\gamma}$ as sets of coefficients.

As a result, Equation (23) leads to:

$$\boldsymbol{\beta} = \langle \mathbf{R},\mathbf{L}\rangle^{-1}\ \langle \mathbf{R},\mathbf{N}\rangle\boldsymbol{\delta}\,. \tag{29}$$

The two bilinear forms in Equation (21) are always symmetric. For the second one, we have:

$$\boldsymbol{\delta}^T\langle \mathbf{R},\mathbf{N}\rangle^T\ \langle \mathbf{R},\mathbf{L}\rangle^{-1}\ \langle \mathbf{R},\mathbf{N}\rangle\boldsymbol{\gamma}\,. \tag{30}$$

We start the algorithm with the initial formulation for $T\left[u_B^+\right]$. The next step is to produce u_B^+. In order to do this, we use the boundary integral equation on Γ_0, derived directly from Green's formula:

$$\left(\frac{1}{2}\,\mathrm{I} + \mathbf{K}\right)\, u_B^+ = \mathbf{V}\, T\left[u_B^+\right] \quad \text{on } \Gamma_0, \tag{31}$$

I denotes the identity operator, **K** and **V** are singular, respectively regular integral operators. In the numerical implementation, however, Equation (31) cannot be solved exactly. This could destroy the symmetry property of Equations (21) and (22). The aim of this paper is to show that it is possible to chose the collocation points independently from the finite element nodes. Because of the large number of collocation points on the boundary of the macro-element, as shown in Figure 1, we get a better accuracy of the stiffness relation, see Equation (30), than with classical coupling techniques [1] without increasing the number of the finite element nodes. As a result, the band width of the whole system will not be enlarged.

NUMERICAL TESTS

We consider a plate with an elliptical cut-out under tension. Because the configuration is symmetric, we have to discretize only a quarter of the structure. In Figure 2 we see a coupling of macro-elements, denoted I and II, with classical displacement finite elements. The question is, in which way the condition number of the influence matrix $\langle R,L\rangle$ depends on the shape of these macro-elements. Several numerical tests have shown that the side lengths d_i (i = 1 to 5) must be nearly the same, see Figure 3. With this shape, we have the best condition number for the kernel matrix $\langle \mathbf{R},\mathbf{L}\rangle$. We see in Figure 4 a test with two macro-elements where the shape is controlled by the parameters x_1 and y_1, see Feurer [10]. In this configuration we see that the side lengths for each macro-element are very different from the shape of a macro-element as shown in Figure 3. Figure 5 shows the condition number as a function of the parameter y_1. We see that if the side lengths are almost equal, we have the smallest condition number, around 300. For that case y_1 has a value of about 1. If y_1 decreases to very small

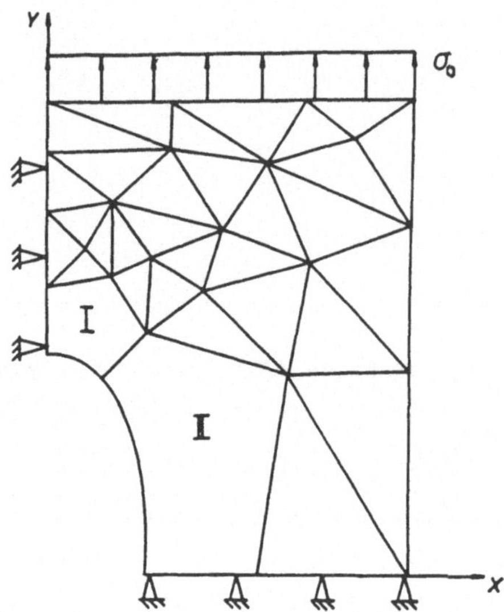

Fig. 2. Discretization of the structure by 2 macro-elements (I, II) and adjacent classical displacement finite elements for a notch problem with an eliptical cut-out.

values, the condition number increases to high values. If y_1 is smaller than 0.1, the condition number is so bad that it is impossible to produce the inverse of the matrix <R,L> on the computer. This applies also to Figure 6 where the condition number is changing with the parameter x_1 for the macro-element II, Figure 2. From x_1 = 0.5 up to 1.3, we have a condition number around 300. Yet, if the value x_1 is smaller than 0.5, the condition number also increases rapidly. If x_1 is less than 0.2, we have no chance to

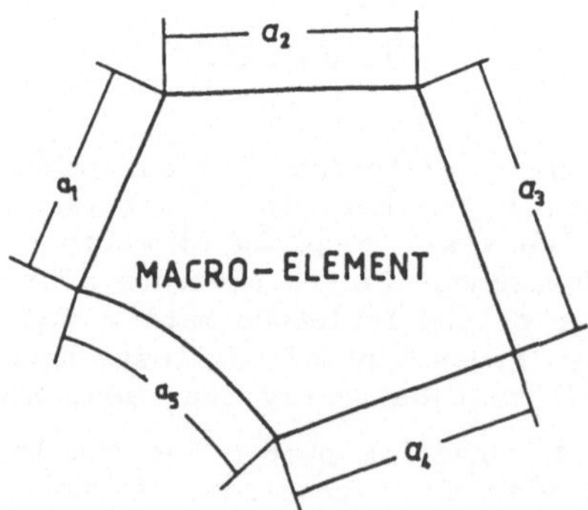

Fig. 3. The best shape for a macro-element leading to the best condition number of the kernel matrix from the stiffness relation.

inverse <R,L> on the computer. This shows that a macro-element with almost equal side lengths (Figure 3) leads to the best condition number. Because

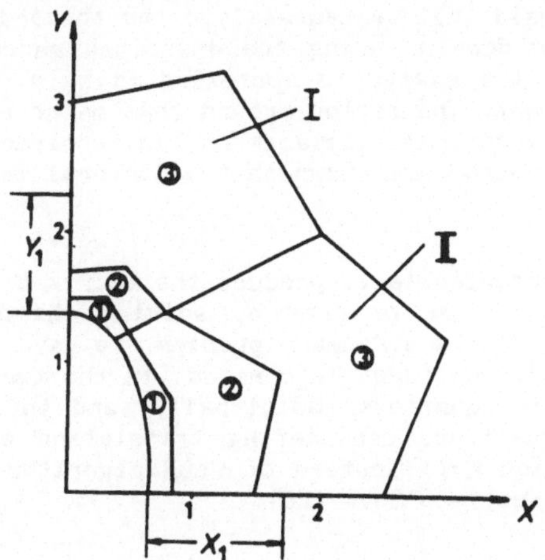

Fig. 4. Several shapes (①, ②, ③ for the macro-elements I and II) The parameters x_1 and y_1 are controlling the shape of the macro-element.

of the great freedom we have in defining macro-elements in the whole structure, there is no problem to choose a discretization that leads to an influence matrix with a low condition number.

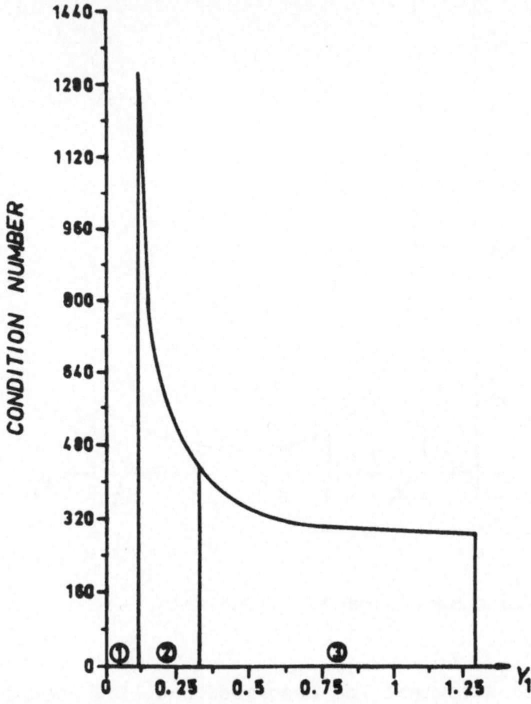

Fig. 5. The condition number depends on the parameters y_1 for the macro-element I

As a result it can be said that we can analyze the three-dimensional cavity problem in an unbounded domain, using the described macro-element technique. The region around the cavity is approximated by a large finite element, called macro-element. The region around that macro-element is discretized by classical tetrahedronal displacement finite elements. Accurate results obtained by this method are shown in the Doctoral Thesis of Karaosmanoglu [14].

Another very important problem is to produce the matrix L in Equation (29). As explained before, this can be done by solving the integral equation (31). Because Equation (31) is a Neumann problem, we have to consider for a unique solution rigid body motions. This means, in the two-dimensional case we have to require three equations additionally, and in three-dimensional case six additional equations, considering translation and rotation. The best is to solve Equation (31) instead of Gauß algorithm to do it by the QR-decomposition technique, see Soyk [20].

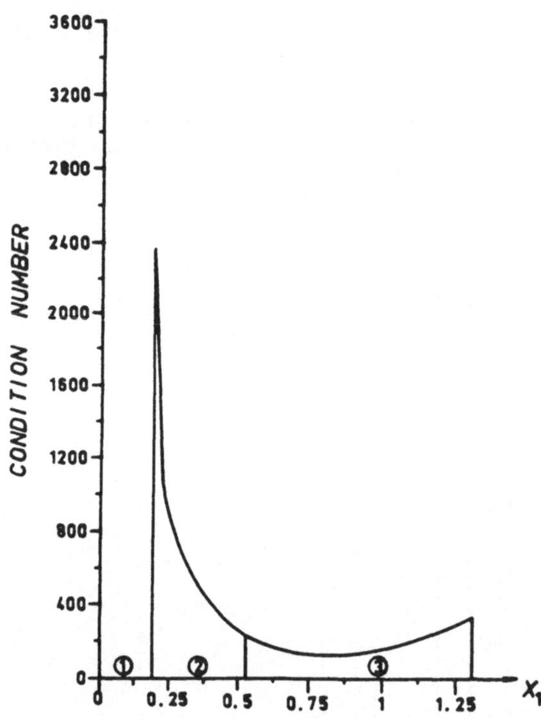

Fig. 6. The condition number depends on parameter x_1 for the macro-element II

All in all it can be said that the described coupling procedure yields accurate results for two- and three-dimensional elasticity problems. The major advantage compared to working with pure FEM or pure BEM consists in the lower CPU-time required for solving the problem.

REFERENCES

[1] Atluri, S.N.; Grannell, J.J. (1978): Boundary Element Methods (BEM) and Combination of BEM-FEM. Report No. GIT-ESM-SA-78-16.

[2] Bart, J. (1974): Kombination eines Integralgleichungsverfahrens mit der FEM zur Berechnung ebener Spannungskonzentrationsprobleme. Doctoral Thesis, University of Munich.

[3] Beer, G. (1983): Finite Element, Boundary Element and Coupled Analysis of Unbounded Problems in Elastostatics. Int. J. for Num. Meth. in Engng. 19, 567-580.

[4] Beer, G.; Meek, J.L. (1981): The Coupling of Boundary and Finite Element Methods for Infinite Domain Problems in Elasto-Plasticity. In: Brebbia, C.A. (Ed.): Boundary Element Methods, Vol. 3, pp. 575-591. New York: Springer-Verlag, Berlin, Heidelberg.

[5] Carmine, R. (1989): Ein Kopplungsverfahren von FEM und BEM zur Berechnung ebener Spannungskonzentrationsprobleme. Doctoral Thesis, Karlsruhe University.

[6] Chen, Z.-S.; Hofstetter, G.; Li, Z.-K.; Mang, H.A.; Torzicky, P. (1990): Coupling of FE- and BE-Discretization for 3D-Stress Analysis of Tunnels in Layered Anisotropic Rock. In: Kuhn, G.; Mang, H. (Eds.): Proceedings of IUTAM/IACM Symposium on Discretization Methods in Structural Mechanics, Vienna/Austria 1989, pp. 427-436. Berlin, Heidelberg: Springer-Verlag.

[7] Cheng, K.C. (1983): Ein gemischtes Verfahren auf der Basis von FEM/BEM. Master Thesis, Report No. 6 of the Institute of Solid Mechanics, Karlsruhe University.

[8] Choi, C.K.; Ahn, J.S. (1988): A Coupled Formulation of Finite and Boundary Element Methods in Two-Dimensional Elasticity. In: Brebbia, C.A. (Ed.): Boundary Elements X, Vol. 1, pp. 477-491, Proceedings of the 10th BEM Conference, Southampton University, 1988. Heidelberg, New York: Springer-Verlag.

[9] Costabel, M. (1987): Symmetric Method for the Coupling of Finite Elements and Boundary Elements. In: Brebbia, C.A.; Wendland, W.L.; Kuhn, G. (Eds.): Boundary Elements IX, Vol. 1, pp. 411-420, Proceedings of the 9th BEM Conference, University of Stuttgart, 1987. Berlin, Heidelberg, New York: Springer Verlag.

[10] Feurer, G. (1987): Einflüsse auf die Kondition der H-Matrix beim Coupling-Verfahren. Report No. 43 of the Institute of Solid Mechanics, Karlsruhe University.

[11] Histake, M.; Ito, T.; Ueda, H (1983): Three Dimensional Symmetric Coupling of Boundary and Finite Element Methods. In: Brebbia, C.A. (Ed.): Boundary Elements 3, pp. 985-994. Berlin, Heidelberg, New York: Springer-Verlag.

[12] Hsiao, G.C. (1988): The Coupling of BEM and FEM. A Brief Review. In: Brebbia, C.A. (Ed.): Boundary Elements X, Vol. 1, pp. 431-446, Proceedings of the 10th BEM Conference, Southampton University, 1988. Ber-

lin, Heidelberg, New York: Springer-Verlag.

[13] Hsiao, G.C.: The Coupling of Boundary Element and Finite Element Methods, invited paper, Proceedings of the GAMM Conference 1989 in Karlsruhe, West-Germany.
to appear in ZAMM

[14] Karaosmanoglu, N. (1989): Kopplung von Randelement- und Finite-Element-Verfahren für dreidimensionale elastische Strukturen. Doctoral Thesis, Karslruhe University.

[15] Mang, A.H.; Chen, Z.Y. (1986): Zur Symmetrisierbarkeit von Kopplungsmatrizen bei FE-BE Diskretisierung fester Körper, private information.

[16] Mustoe, G.G.W. (1979): Coupling of Boundary Solution Procedures and Finite Elements for Continuum Problems. Doctoral Thesis, University of Wales, University of Swansea.

[17] Schnack, E. (1973): Beitrag zur Berechnung rotationssymmetrischer Spannungskonzentrationsprobleme mit der Methode der finiten Elemente. Doctoral Thesis, University of Munich.

[18] Schnack, E. (1987): A Hybrid BEM Model. J. Num. Meth. Engng. 24, 1015-1025.

[19] Schnack, E. (1985): Stress Analysis with a Combination of HSM and BEM. In: Whiteman, J.R. (Ed.): Mathematics of Finite Elements and Application V, pp. 273-281, Proceedings of the Mafelap Conference, Brunel University, 1984. London, New York, Tokyo: Academic Press, Inc.

[20] Soyk, W.. (1990): Untersuchung zur Symmetrie der H-Matrix bei der FEM-BEM-Kopplung. Report No. 62 of the Institute of Solid Mechanics, Karlsruhe University.

[21] Wendland, W.L. (1988): On Asymptotic Error Estimates for Combined BEM and FEM. In: Stein, E.; Wendland, W.L. (Eds.): Finite Element and Boundary Element Techniques from Mathematical and Engineering Points of View, CISM Courses and Lectures, No. 301, pp. 273-333. Berlin, Heidelberg, New York: Springer-Verlag.

[22] Wendland, W.L. (1990): On the Coupling of Finite Elements and Boundary Elements. In: Kuhn, G.; Mang, H. (Eds): pp. 405-414, Proceedings of the IUTAM/IACM Symposium on Discretization Methods in Structural Mechanics, Vienna, Austria, 1989, pp. 405-414. Berlin, Heidelberg: Springer-Verlag.

[23] Zienkiewicz, O.G.; Kelly, D.W.; Bettess, P. (1979): Marriage a la Mode - The Best of both Worlds (Finite Elements and Boundary Integrals). In: Glowinky, R.; Rodin, E.Y.; Zienkiewics, O.C. (Eds.): Engineering Methods in Finite Element Analysis, Chapter 5, pp. 81-105. Chichester, New York, Brisbane and Toronto: John Wiley and Sons.

ON THE NUMERICAL INTEGRATION OF SINGULAR SURFACE INTEGRALS IN THE BEM

C. Schwab
Dept. of Mathematics and Statistics, The University of Maryland,
Baltimore County Campus, Baltimore, MD 21228-5398, U.S.A.

W.L. Wendland
Mathematisches Institut A, Universität Stuttgart,
Pfaffenwaldring 57, D-W-7000 Stuttgart 80, Germany.

SUMMARY

The properties of pseudo-differential operators allow a unified and detailed analysis of numerical integration. We present error estimates exploiting the Rabinowitz-Richter estimates for three numerical integration techniques of boundary integral operators on an analytic boundary element containing the source point. We first study weakly singular integral operators which are pseudo-differential operators of order ≤ -1. We show that Duffy's triangular coordinates transform all these operators to regular integrals and provide asymptotic error estimates for corresponding Gaussian quadrature. Similar results can be obtained for plane polar coordinates in the parameter domain with scaled Gaussian quadrature in radial and Gaussian quadrature on the angular pieces of analyticity. We also find that Lyness extrapolation for all weakly singular boundary integral operators is asymptotically almost as efficient.
For strongly singular integral operators defined by Cauchy singular and Hadamared finite part integrals which are pseudo-differential operators of orders ≥ 0 we analyze product integration with Kutt formulas in radial and Gaussian integration on the angular pieces of analyticity. Again, we give asymptotic error estimates.

Acknowlegement: This work was partly supported by the Priority Research Programme "Boundary Element Methods" of the German Research Foundation DFG under Grant Nb. We 659/16-1.

1. INTRODUCTION

In the last decade the boundary element method became a reliable and efficient tool for the numerical solution of boundary value problems [16]. The method is based on a reduction of the original problem in $\mathbb{R}^3$ to an integral equation on the boundary surface Γ. The subsequent discretization of this integral equation requires the numerical evaluation of weakly singular, Cauchy singular and, more recently, in connection with FEM-BEM coupling procedures and crack- problems (see. e.g. [17] and [4]), of hypersingular integrals over general, curved boundary element patches.
All boundary integral operators arising in regular elliptic boundary value problems are, quite generally, strongly elliptic pseudo-differential operators of integer order n (see. e.g. [2]) which are applied to trial functions with small support on Γ. We address in this article the case that the source point x, which is either a collocation point or an integration knot for the outer integration of a Galerkin scheme, lies in the support of the trial function. We present and analyze numerical integration schemes, some of which have been sucessfully used in applications, in the context of h-boundary elements

where convergence is achieved by reducing the meshwidth h. Our error estimates allow to assess a-priori the number of integration knots required to guarantee any prescribed consistency error for the singular integrals as h tends to zero. Moreover, the framework of pseudo-differential operators allows us to develop integration schemes that work for a whole class of boundary integrals, rather than only for one specific operator. So we prove, for example, that Duffy's triangular coordinates are applicable to all weakly singular boundary integrals and we develop an integration scheme, based on local polar-coordinates that can be used for weakly singular as well as for Cauchy and hypersingular operators.

The plan of this article is as follows: in Section 2 we present the assumptions on the boundary surface Γ and on our boundary elements that are used in the subsequent analysis. In addition, we present results on the representation of general boundary integral operators in local coordinates including a definition of finite part and Cauchy singular integrals.
In Section 3 we treat the numerical quadrature of weakly singular integrals. Here we obtain error estimates for Duffy's triangular coordinates [3], for a modified extrapolation scheme due to Lyness [8], [9] and for a quadrature which is based on polar coordinates in the parameter domain. All these methods are shown to work for any weakly singular boundary integral operator, in particular for the single- and double layer potentials.
In Section 4 we consider the quadrature of Cauchy and hypersingular integrals. Here we analyze tensor-product integration formulas in local polar coordinates. Finally some examples are presented which illustrate the construction of the new formulas.
We point out that here we mainly present our results; the proofs of all assertions below can be found in [13] and [14].

2. BOUNDARY INTEGRAL OPERATORS AND SINGULAR SURFACE INTEGRALS

Let $\Omega \subset \mathbb{R}^3$ be an interior bounded domain with piecewise analytic boundary $\Gamma = \partial\Omega$, i.e. Γ is partitioned into finitely many pieces S_l so that $\Gamma = \cup \overline{S_l}$ and each S_l is the image of a parameter domain $V_l \subset \mathbb{R}^2$ under an analytic mapping $y = \chi_l(v_1, v_2)$ that has an analytic extension beyond $\overline{V_l}$. In addition, we require Γ to be a closed Lipschitz surface, see e.g.[11].
As was pointed out in the introduction, every boundary integral operator arising in BEM for elliptic problems can be represented by a (matrix of) classical pseudo-differential operator(s) of integer order(s). In the local coordinates $(v_1, v_2) = v$, any such operator applied to a function $u \circ X_l^{-1}$ having compact support on S_l can be written in the form

$$\begin{aligned}(B_n u)(v) \;=\; & \sum_{|\lambda|\leq n} b_\lambda(v)(D^\lambda u)(v) \\ & + \text{p.f.} \int_{s\in\mathbb{R}^2} \left\{ \frac{1}{r^{n+2}} \sum_{j=0}^{L} f_{n-j}(v,\theta) r^j + \ln r \sum_{j=0}^{L-n-1} f^*_{n-j}(v,\theta) r^j \right\} u(s)ds \\ & + R_L u \end{aligned} \tag{1}$$

where the so-called characteristics $f_\kappa(v,\theta)$ satisfy for $\kappa \leq -2$ and $\kappa \geq 0$ the *compatibility conditions:*

$$\int_{\theta=0}^{2\pi} f_\kappa(v,\theta) \cos^{\alpha_1}\theta \sin^{\alpha_2}\theta \, d\theta = 0 \text{ for} \begin{cases} |\alpha| = \kappa & \text{if} \quad \kappa \geq 0, \\ |\alpha| = -\kappa - 2 & \text{if} \quad \kappa \leq -2. \end{cases} \tag{2}$$

$r = |v - s|$ and θ denote polar-coordinates in V_l at v. $R_L u$ denotes the remainder given by an operator of order $n - L - 1$. In (1) p.f.denotes the Hadamard finite part which is defined as follows:
For $\varepsilon > 0$ and $\kappa \geq 0$ consider

$$I_\varepsilon^\kappa[u] := \int_{s \in \mathbb{R}^2 \setminus \{|v-s| < \varepsilon\}} r^{-2-\kappa} f_\kappa(v, \theta) u(s) ds.$$

If we let $\varepsilon \to 0$, we obtain the asymptotics

$$I_\varepsilon^\kappa[u] \sim C_0 \log \varepsilon + \sum_{j=1}^{\kappa} C_j \varepsilon^{-j} + \text{finite part}(I_\varepsilon^\kappa). \tag{3}$$

Following Hadamard, we discard the divergent terms and take the remaining finite part to be the value of the integral. This leads to the following expression where a Taylor polynomial of u at v has been subtracted so that only regular integrals appear:

Definition 1

$$\begin{aligned}
&\text{p.f.} \int_{\Omega_0} \frac{f_\kappa(v, \theta)}{r^{2+\kappa}} u(s) ds \\
&= \sum_{|\alpha| \leq \kappa} \frac{1}{\alpha!} D^\alpha u(v) \int_0^\omega f_\kappa(v, \theta) \cos^{\alpha_1} \theta \sin^{\alpha_2} \theta \left\{ \begin{array}{ll} \ln R(\theta) & \text{for } |\alpha| = \kappa \\ (|\alpha| - \kappa)^{-1} R^{|\alpha| - \kappa}(\theta) & \text{for } |\alpha| < \kappa \end{array} \right\} d\theta \\
&+ \int_0^\omega \int_0^{R(\theta)} r^{-2-\kappa} f_\kappa(v, \theta) \left\{ u(s) - \sum_{|\alpha| \leq \kappa} \frac{1}{\alpha!} (D^\alpha u(v))(s - v)^\alpha \right\} ds.
\end{aligned}$$

Here $R(\theta)$ is a parametrization of $\partial(\text{conv}(\text{supp} u))$. If the characteristic f_κ in $I_\varepsilon^\kappa[u]$ satisfies certain compatibility conditions then the constants C_j in (3) vanish and the integral exists in the "Cauchy principal-value" sence (see [13] for details). It is well known that both, principal value and finite part integrals are *not* invariant unter smooth changes of variables. What is in fact invariant that is the form (1), i.e. the finite part integral together with a differential operator of order n. Changing variables in I_ε^κ results in a change of the coefficients $b_\lambda(v)$ in (1) and the precise dependence is given in [13, Theorem 2] .
Let us first show how the singular surface integrals ordinarily encountered in BEM fit into the framework (1). We assume that the triangulations on Γ are images of regular partitions of V_l with maximum meshwidth h under the parameter representations χ_l. Regular here means that all elements in V_l are obtained by a translation t_j and one linear mapping A_l from a scaled master element $\Omega_{0l}^h = h\Omega_{0l}$ of size $O(h^2)$. Consequently, we have for our boundary patches $F_{jl}^h \subset S_l$:

$$F_{jl}^h = (\chi_l \circ A_l)(\Omega_{0l}^h + t_j) =: \overline{\chi}_{jl}(\Omega_{0l}^h). \tag{4}$$

Therefore the surface representation is exact and the parametrization of S_l is independent of h. This allows to obtain analytic extensions of the transformed BEM kernels with extension domains independent of h which is used in our error estimates below. The integrals to be evaluated are of the form

$$I^h(x) = \text{p.f.} \int_{y \in F^h} K_n(x, y - x) u(y) ds(y) \tag{5}$$

where F^h is any of the patches in (4) and K_n is the kernel associated with a boundary integral operator of order n, such as e.g. in potential theory:

$$n=-1: \quad K_n(x,y-x) = |y-x|^{-1} \text{ (single layer potential)}, \tag{6}$$

$$K_n(x,y-x) = \frac{\vec{\nu}(y)\cdot(y-x)}{|y-x|^3} \text{ (double layer potential)}, \tag{7}$$

$$n=0: \quad K_n(x,y-x) = \frac{\vec{a}(y)\cdot(y-x)}{|y-x|^3} \text{ (oblique derivative kernel)}, \tag{8}$$

$$n=1: \quad K_n(x,y-x) = |y-x|^{-3} \text{ (hypersingular potential)}. \tag{9}$$

All kernels K_n are homogeneous, i.e. they satisfiy

$$K_n(x,tz) = t^{-2-n}K_n(x,z) \quad \text{for all} \quad t>0.$$

Hence (5) is a Hadamard finite part integral for $n \geq 0$ and for $n = 0$ even a Cauchy principal value integral.
By using Taylor expansion of χ_l, Guiggiani [5],[6] computes explicitly the kernel expansions corresponding to Definition 1 for the corresponding operators of elasticity.
To evaluate $I^h(x)$ in (5) numerically, we use the parameter representation $\overline{\chi}$ in (4) and rewrite I^h as an integral over the reference element Ω_0 (the subscribs of χ_l have been dropped here). For weakly singular operators this does not cause any difficulties, whereas for $n \geq 0$ we have

Theorem 1(Kieser [7]):
If $K_n(x,y-x)$ in (5) corresponds to a boundary integral operator associated with a regular elliptic boundary value problem, then the classical rule of variable substitution is valid.

Consequently, if $x = \overline{\chi}(u), y = \overline{\chi}(v)$ we may write (5) as

$$\tilde{I}^h(u) = \int_{\Omega_0^h} \tilde{K}_n(v-u)\psi(v)dv \tag{10}$$

where now, however, $\tilde{K}_n$ is only pseudo-homogeneous, i.e.

$$\tilde{K}_n(v-u) \sim \sum_{j\geq 0} \tilde{K}_{n,j}(v-u) \qquad \text{as} \qquad v \to u \tag{11}$$

where each $\tilde{K}_{n,j}$ is homogeneous of degree $n-2+j$. The function ψ in (10) is a smooth density containing the surface element $|ds(y)|$. We see that the integral in (10) is just a special case of (1) with the characteristics depending on derivatives of $\overline{\chi}$ at u.
Remark 1: The Rabinowitz-Richter estimates will require an estimate of the domain of analyticity, associated with the analytic continuation of the integrands. Hence, we need the domains of analyticity with respect to $\theta \in \mathbb{C}$. To obtain the functions in the respresentation in Definition 1 from $K(x;y-x)$, we insert the analytic parameter representation in local polar coordinates

$$y = \chi(u + r(\cos\theta, \sin\theta)^T).$$

Clearly, the analyticity of χ implies convergence of the power series expansions with respect to r and θ, respectively, in a common fixed domain $\mathcal{U} \subset \mathbb{C}^2, (r,\theta) \in \mathcal{U} \supset [0,r^*]\times\mathbb{R}$ where $r^* > 0$ and $\mathcal{U}$ are independent of the partitions of the parameter domains and

—by Heine-Borel's theorem— independent of u. The relation (11) is a consequence of the following Theorem which also sheds light on the structure of the characteristics f_{n-j} in (1):

Theorem 2([13, Theorem 4]):
Let $\rho = |x-y|$ and $r = |u-v|$. Then, for every integer α we have

$$\rho^{-2-\alpha} = r^{-2-\alpha} \sum_{\nu=0}^{\infty} (l_2^{-\frac{\alpha}{2}-\nu-1} P_{3\nu})(\cos\theta, \sin\theta) r^{\nu}. \tag{12}$$

Here $P_{3\nu}$ denotes a homogeneous polynomial of degree 3ν in $\sin\theta$ and $\cos\theta$ and

$$l_2(\theta) = |\overline{\chi}_{|1}||\overline{\chi}_{|2}|\{\lambda\cos^2\theta + 2\cos\gamma\cos\theta\sin\theta + \lambda^{-1}\sin^2\theta\} \tag{13}$$

with

$$\lambda = |\overline{\chi}_{|1}|/|\overline{\chi}_{|2}| \tag{14}$$

and

$$\cos\gamma = \frac{\overline{\chi}_{|1}\cdot\overline{\chi}_{|2}}{|\overline{\chi}_{|1}||\overline{\chi}_{|2}|}. \tag{15}$$

Remark 2:
The $\overline{\chi}_{|1}, \overline{\chi}_{|2}$ are the tangent vectors to the surface in x; and γ is the angle between them. The quadratic form (13) is the first fundamental form in $\sin\theta$ and $\cos\theta$ depending only on λ and $\cos\gamma$, i.e.

$$l_2(\theta) = |\overline{\chi}_{|1}||\overline{\chi}_{|2}| \begin{pmatrix} \cos\theta \\ \sin\theta \end{pmatrix}^T \begin{pmatrix} \lambda & \cos\gamma \\ \cos\gamma & \frac{1}{\lambda} \end{pmatrix} \begin{pmatrix} \cos\theta \\ \sin\theta \end{pmatrix}.$$

The eigenvalues of the matrix in l_2 are

$$\mu = \frac{1}{2}(\lambda + \frac{1}{\lambda}) \pm \sqrt{\frac{(\lambda+\lambda^{-1})^2}{4} - \sin^2\gamma}.$$

Hence, for all γ between 0 and π we have $0 < \mu < \lambda + \lambda^{-1}$, implying that the form is positive definite and l_2 does not vanish for all $\theta \in [0, 2\pi]$. Since all the kernels K_n in (6)–(9) are of the form

$$K_n = \left\{ \sum_{|\alpha|\geq\nu} a_\alpha(u) v^\alpha \right\} \rho^{-\nu-n-2} \tag{16}$$

with some $\nu \geq 0$ (e.g. in (6) we have $\nu = 0$, in (7) $\nu = 2$ and in (8) $\nu = 1$, in (9) $\nu = 0$). Theorem 2 gives rather precise information on the structure of $\tilde{K}_n$, the transformed kernel (11); and this is the basis for the design and analysis of efficient quadrature schemes for (10). We emphasize that the integrand in (10) is independent of h.

3. NUMERICAL INTEGRATION OF WEAKLY SINGULAR INTEGRALS ($n = -1$)

In this section we will analyze Duffy's triangular coordinates [3], an extrapolation method due to Lyness [8],[9] and a tensor product approach based on the polar-coordinates.

3.1 Duffy's triangular coordinates

For simplicity, we assume $u = 0$ in (11) and consider

$$\tilde{I}^h = \int_{\Omega_0^h} \tilde{K}_{-1}(v_1, v_2)\psi(v_1, v_2)dv_1 dv_2$$

where

$$\tilde{K}_{-1}(v) = \left\{ \sum_{|\alpha|\geq\nu} a_\alpha(u)v^\alpha \right\} \rho^{-\nu-1} \circ \chi_\ell \tag{17}$$

and Ω_0^h is as in Fig.1:

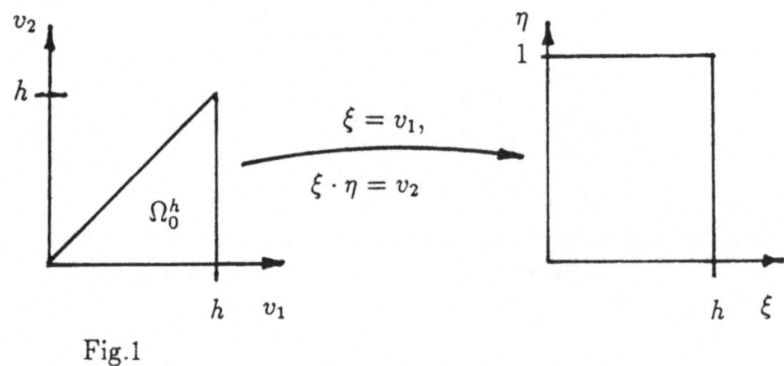

Fig.1

We introduce Duffy's triangular coordinates

$$\xi = v_1, \qquad \xi \cdot \eta = v_2 \tag{18}$$

and have

Theorem 3:
Let χ be an analytic parametric representation of Γ. Then substituting (18) into the integrand (17) yields an integrand which is complex analytic in ξ for fixed η and admits an analytic continuation in η up to $\pm i$ and to $\lambda e^{\pm i\gamma}$, respectively. Here λ, γ are as in (13) and are independent of ξ and h, provided, the meshwidth h is smaller than r^* in Remark 1.

Proof: From (18) we have

$$dv = \xi\, d\xi\, d\eta, \cos\theta = \xi\, r^{-1} \text{and } \sin\theta = \xi\,\eta\, r^{-1}.$$

Inserting these expressions into the right hand side of (17) and using Theorem 2 for $\alpha = -1$, we see that the integrand becomes

$$\tilde{K}_{-1}(\xi, \xi\eta)\, dv = \sum_{j=0}^{\infty} \sum_{\mu=0}^{\infty} \xi^{j+\mu} p_{3\mu}(1,\eta) Q_{j+\nu}(1,\eta) \left(c\left\{ \lambda + 2\cos\gamma\eta + \lambda^{-1}\eta^2 \right\} \right)^{-\frac{\nu}{2}-\mu-\frac{1}{2}} d\xi\, d\eta\,. \tag{19}$$

3. NUMERICAL INTEGRATION OF WEAKLY SINGULAR INTEGRALS ($n = -1$)

In this section we will analyze Duffy's triangular coordinates [3], an extrapolation method due to Lyness [8],[9] and a tensor product approach based on the polar-coordinates.

3.1 Duffy's triangular coordinates

For simplicity, we assume $u = 0$ in (11) and consider

$$\tilde{I}^h = \int_{\Omega_0^h} \tilde{K}_{-1}(v_1, v_2)\psi(v_1, v_2)dv_1dv_2$$

where

$$\tilde{K}_{-1}(v) = \left\{ \sum_{|\alpha|\geq\nu} a_\alpha(u)v^\alpha \right\} \rho^{-\nu-1} \circ \chi_\ell \tag{17}$$

and Ω_0^h is as in Fig.1:

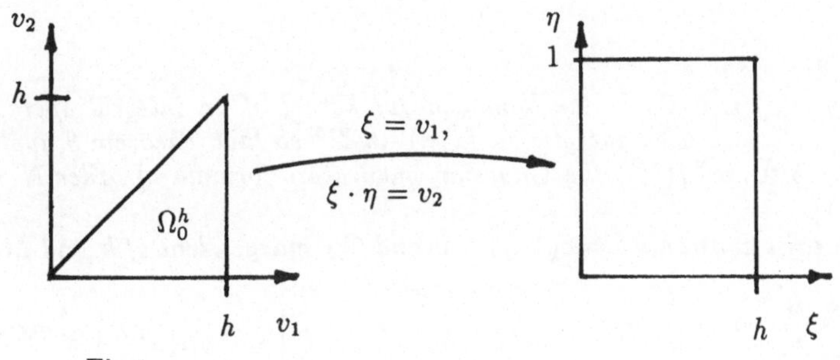

Fig.1

We introduce Duffy's triangular coordinates

$$\xi = v_1, \qquad \xi \cdot \eta = v_2 \tag{18}$$

and have

Theorem 3:
Let χ be an analytic parametric representation of Γ. Then substituting (18) into the integrand (17) yields an integrand which is complex analytic in ξ for fixed η and admits an analytic continuation in η up to $\pm i$ and to $\lambda e^{\pm i\gamma}$, respectively. Here λ, γ are as in (13) and are independent of ξ and h, provided, the meshwidth h is smaller than r^* in Remark 1.

Proof: From (18) we have

$$dv = \xi\, d\xi\, d\eta, \cos\theta = \xi\, r^{-1} \text{and } \sin\theta = \xi\,\eta\, r^{-1}.$$

Inserting these expressions into the right hand side of (17) and using Theorem 2 for $\alpha = -1$, we see that the integrand becomes

$$\tilde{K}_{-1}(\xi, \xi\eta)\, dv =$$
$$\sum_{j=0}^{\infty} \sum_{\mu=0}^{\infty} \xi^{j+\mu} p_{3\mu}(1,\eta) Q_{j+\nu}(1,\eta) \left(c\left\{\lambda + 2\cos\gamma\eta + \lambda^{-1}\eta^2\right\}\right)^{-\frac{\nu}{2}-\mu-\frac{1}{2}} d\xi\, d\eta\,. \tag{19}$$

If χ is analytic in a neighborhood of x, then (19) is complex analytic in ξ in some neighborhood of $(0,0)$, and the series converges.
The only term which may account for a loss of analyticity in η is $\{\lambda+2\cos\gamma\eta+\lambda^{-1}\eta^2\}^2$, which fails to be analytic if

$$\lambda + 2\cos\gamma\eta + \lambda^{-1}\eta^2 = 0$$

or

$$\eta_{1,2} = \lambda\cos\gamma \pm \sqrt{\lambda^2\cos^2\gamma - \lambda^2} = \lambda(\cos\gamma \pm i\sin\gamma) = \lambda e^{\pm i\gamma}$$

as proposed in the theorem.
The points $\eta = \pm i$ need to be excluded in order to assure the convergence of the series (19). Consequently, we may use Gaussian quadrature for the integration of (19). Since the domain of integration is as indicated in Fig.1, however, we can obtain error estimates in powers of h only for the ξ–integration with a standard scaling argument.
The error in the η–integration must be estimated differently, using the estimates [12]. We have

Theorem 4:
Let $K_n(x, y-x), n \in \mathbb{Z}$ be the homogeneous kernel of an integral operator of order $n \leq -1$ on Γ, a piecewise analytic surface Γ in $\mathbb{R}^3$ so that Theorem 3 applies. Denote by $G_\xi^N(f) = \sum_{j=1}^N w_j^N f(\xi_j^N)$ the Gaussian quadrature formula of order N in ξ for the interval [0,1].
Then there exist constants $\delta = \delta(\lambda,\gamma) > 0$ and C_N independent of h and M, such that

$$\left|\sum_{i=1}^{N}\sum_{j=1}^{M} w_i^N w_j^M [h\xi_i^N(\tilde{K}_n\psi)](h\xi_i^N, h\xi_i^N\eta_j^M) - I_{jl}^h\right| \leq C_N h\{h^{2N} + e^{-2M\ln\delta}\}$$

where

$$I_{jl}^h = \int_{F_{jl}^h} K_n(x, y-x)\psi \circ \chi_l^{-1}(y)ds(y).$$

The quantity δ can be obtained explicitly in terms of λ and γ introduced in (13).

Remark 3: We always have $1 \leq \delta \leq 3.58\cdots$. In particular, if a family of boundary partitions is regular with $0 < \gamma_* < \gamma$ and $\gamma + \frac{1}{\gamma} \leq \Lambda$ where the constants γ_* and Λ are fixed for the whole family then $\delta_* = \delta(\gamma_*, \Lambda)$ is a lower bound for all δ of the family, and can be used in conjunction with estimate (20) below to select the number of integration knots necessary to guarantee consistency as the mesh is refined. The proof can be found in [14].

Corollary
In order to satisfy the consistency estimate

$$|E| \leq Ch^a$$

for the integration error E for the numerical integration in Theorem 4 and for given order of consistency $a > 0$, one should choose

$$N \geq \frac{a-1}{2} \text{ and } M \geq \frac{|\ln h|}{\ln\delta}N \geq \frac{a-1}{2}\frac{|\ln h|}{\ln\delta}. \tag{20}$$

3.2 Extrapolation techniques

As is well known, asymptotic expansion of the integration error allows for extrapolation techniques based on regular subdivisions of the domain of integration. For smooth integrands, the classical Romberg extrapolation yields highly accurate results efficiently. J. Lyness has extended extrapolation to multidimensional integration with integrands that have point singularities [8],[9],[10]. We will show that his method applies to all weakly singular kernels in local coordinates.

Theorem 5:
Let $K(x, y-x)|_\Gamma$ define a $\psi do_{-1}(\Gamma)$, where Γ is a (piecewise) sufficiently smooth surface in $\mathbb{R}^3$. If χ is a sufficiently smooth local chart, then

$$(K \circ \chi)(u, v-u) =: \tilde{K}(u, v-u)$$

is in the class $H_{-1}^{(2)}$ introduced in [8, Definition 5.2,Lemma 5.3]. *Correspondingly, the error estimates in* [8, Theorem 5.14] *hold.*

This theorem is an immediate consequence from our kernel-expansions [13, Theorem 4] and the definition of $H_{-1}^{(2)}$ in [8].
Let us now briefly sketch the key ideas of the extrapolation method. We refer to [8] for more details and the notation. The integral to be evaluated is denoted by $I[\tilde{K}\psi]$, with $\tilde{K}\psi \in H_{-1}^{(2)}$, and the domain of integration $(0,1)^2$. By $Q^{(m)}$ we denote the m-copy of a simple (e.g. midpoint) quadrature rule applied to $\tilde{K}\psi$. Then the following holds:
Proposition [8, Theorem 5.14]:

$$Q^{(m)}[\tilde{K}\psi] - I[\tilde{K}\psi] = \sum_{s=1}^{l-1} \frac{A_s + B_s}{m^s} + \frac{C_s}{m^s} \ln m + O(m^{-l} \ln m) \quad \text{with} \quad l \geq 2. \tag{21}$$

In extrapolation the remainder is discarded, an order $l \geq 2$ is chosen and $Q^{(m)}[\tilde{K}\psi]$ is evaluated for $2l-1$ values $1 \leq m_1 < m_2 \cdots < m_{2l-1}$.

Then, unlike in the classical Romberg scheme, the unkown coefficients in the expansion are obtained by solving a linear system of $2l-1$ equations, with the first component of the solution vector being the extrapolated value.
As is usual in extrapolation, the coefficient matrix depends on the sequence of subdivisions of the reference element and on the singularity of the integrand function. We note that in case Ω_0 is the triangle, an analogous result holds as well, see [10]. In practical applications the condition number of the coefficient matrix depends strongly on the sequence $\{m_j\}$ used. If this condition number is required to be small then strong refinements are needed and hence a rather large number of kernel evaluations. In some test calculations we made the following
Observations:

i) The only sequence $\{m_j\}$ of subdivisions yielding competitive results is the harmonic sequence, $\{j\}$.

ii) Due to the ill conditioning of the extrapolation matrix in this case, all calculations should be performed in double precision.

iii) The algebraic overhead (i.e. work involved for the solution of the linear system) is negligible in the case of BEM applications due to "skillful implementation" since the extrapolation coefficient matrix which belongs to one particular type of singularity is independent of the boundarypatch and can be decomposed beforehand once for all.

iv) Under these conditions we achieved a relative error of $\sim 10^{-5}$ for the self elements with at most 200 kernel evalutions.

We now estimate the number of kernel evaluations and the error in the extrapolated value based on the expansion (21) and the harmonic sequence

$$\{m_i\} = \{1, 2, \cdots, 2l-1\}.$$

Since $(0,1)^2$ is subdivided into m^2 subregions and the midpoint rule is applied to each, $Q^{(m)}$ involves m^2 evaluations of $\tilde{K}\psi$. So, for a given order l of the extrapolation, evaluation of the sequence $\{Q^{(m_i)}\tilde{K}\psi\}_{i=1}^{2l-1}$ uses

$$\sum_{s=1}^{2l-1} s^2 = \frac{4}{3}l^3 + O(l^2) =: n_{eval} \tag{22}$$

kernel evaluations. From (21), the error in the extrapolated value is bounded by

$$C(2l-2)^{2-2l}\ln(2l-2),$$

where C is independent of l. Let us assume $C = O(1)$ and find l from the consistency requirement for BEM

$$(2l-2)^{2-2l}\ln(2l-2) \le h^a. \tag{23}$$

Clearly $l = l(a,h)$ and from (23) it follows that

$$\ln l + \ln\{\ln(2l-1) - \frac{1}{l}\ln\ln(2l-1)\} \ge \ln a + \ln\ln(h^{-1}).$$

This is certainly satisfied for $l \ge l^*(a,h)$ where $\ln l^* = \ln a + \ln\ln(h^{-1})$. Consequently,

$$l \ge a\ln(h^{-1})$$

and, from (22) we can now estimate how many kernel evaluations are needed to ensure (23) as $h \to 0$:

$$\begin{aligned} n_{eval} &= \frac{4}{3}l^3 + \text{ lower order terms} \\ &\ge \frac{4}{3}(a|\ln h|)^3 + \text{ lower order terms.} \end{aligned}$$

Let us compare this work estimate to that for triangular coordinates. In the Corollary we found that asymptotically

$$n_{eval} \doteq \frac{a-1}{\ln\delta}|\ln h|$$

evaluations were needed to ensure the order h^a for the integration error. Therefore extrapolation is asymptotically less efficient than product integration based on triangular coordinates.

3.3 Polar coordinates

Consider again

$$\tilde{I}^h = \int_{\Omega_0^h} \tilde{K}_n(v_1, v_2)\psi(v_1, v_2)dv_1dv_2.$$

We set $v_1 \doteq r\cos\theta, v_2 = r\sin\theta$ and get

$$\tilde{I}^h = \int_{\theta_1}^{\theta_2}\left(\int_{r=0}^{hR(\theta)} r\cdot\tilde{K}_n(r\cos\theta, r\sin\theta)\psi(r\cos\theta, r\sin\theta)dr\right)d\theta.$$

Since $n = -1$, it is clear that the factor r cancels the singularity. Moreover, the integrand thus obtained can be shown to be analytic in r and θ. Denoting by G_θ^M the M-point Gaussian quadrature formula for the θ-integration and by G_r^N the N-point Gaussian quadrature with respect to r, we have:

Theorem 6:
There exists C_N independent of h and M such that

$$|(\tilde{I}^h - G_\theta^M G_r^N)[r\tilde{K}_n \psi]| \leq C_N h\{h^{2N} + e^{-2M\ln\delta}\}.$$

Therefore the accuracy that can be achieved with polar-coordinates is asymptotically equal to that for triangular coordinates. We point out, however, that the polar-coordinate approach generalizes to Cauchy- and hypersingular integrals as well, as we will see in Section 4.

3.4 Remarks on Weighted Integration Formulas

In this section we discuss why weighted quadrature formulas for weakly singular kernels are generally not suitable, except if one uses isothermal local coordinates χ. For simplicity we shall focus on the single and double layer potentials (6) and (7), respectively. Let $F \subset \Gamma$ be any boundary patch which is a smooth regular image of the master-element Ω_0^h with

$$F = \chi(\Omega_0^h), x = \chi(u), y = \chi(v).$$

Then the surface integrals over Ω_0^h take the form

$$\int_{\Omega_0^h} \left\{ \frac{1}{\rho \circ \chi} \right\} \psi(v) dv \tag{24}$$

and

$$\int_{\Omega_0^h} \left\{ \frac{\partial}{\partial \nu_y} \left(\frac{1}{\rho} \right) \circ \chi \right\} \psi(v) dv. \tag{25}$$

From the homogeneity we obtain that in both cases

$$\overline{\lim_{v \to u}} |\{\cdots\} r| < \infty, \tag{26}$$

where $\{\cdots\}$ denotes either of the kernels in (24) or (25). Therefore, the following approach for the numerical integration of (24) and (25) is often suggested:
Consider an integration formula Q with weights w_μ and knots $q_\mu \in \Omega_0^h$ for the integral

$$I[g] = \int_{\Omega_0^h} \frac{1}{r} g(v) dv \tag{27}$$

which is exact for all g in the set of polynomials of degree N, i.e. on

$$V = \Pi_N = \text{span}\{v^\alpha | \alpha \in \mathbb{N}_0^2, |\alpha| \leq \text{N}\}. \tag{28}$$

This yields approximations of (24) and (25) of the form

$$Q[r \cdot \{\cdots\}] = \sum_{\mu=1}^{M} w_\mu |u - q_\mu| \cdot \{\cdots\}(q_\mu).$$

The success of such an approach crucially depends on the approximability of $r\{\cdots\}$ by ploynomials in v_1, v_2 i.e. on the smoothness of the limit (26) in Cartesian coordinates.

Howover, the expansion (2) shows that the limit (26) in general is not even in $C^1(\Omega_0^h)$. Sufficient conditions for the boundary parametrization χ such that (26) is smooth, are that $\chi(\cdot)$ be affine and isothermal (i.e. $\lambda = 1$ and $\gamma = \frac{\pi}{2}$) at x. These conditions also seem to be neccessary, since numerical experiments [1] show a rapid loss of accuracy as soon as χ is slightly nonisothermal (i.e. γ in (13) differs slightly from $\frac{\pi}{2}$). In this case, the integrand $\{\cdots\}r$ behaves as $(l_2(\psi))^a$ with some power a and cannot be well approximated by polynomials on Ω_0^h.
Let us als remark that in the case of affine, not isothermal χ, virtually all integrals have been evaluated analytically. We refer to [4],[5],[6] for details.

4. STRONGLY SINGULAR INTEGRALS $n \geq 0$

Here we consider integrals of the form

$$I_{jl}^h = \text{p.f.} \int_{F_{jl}^h} K_n(x, y - x)\psi(\chi_l^{-1}(y))ds(y) \tag{29}$$

where $K_n(x, tz) = t^{-2-n}K_n(x, z)$ for $t \geq 1$ and where p.f. denotes a finite part integral, $x \in F_{jl}^h$ and the order is $n \geq 0$. As before we assume that χ is an analytic representation which maps a domain of the parameter plane to Γ. Correspondingly, meshes on Γ are images of regular triangulations of the parameter plane under χ_ℓ. Clearly, as explained above, F_{jl}^h is, due to our assumption, the image of $\Omega_0^h = \{(0,0),(h,0),(h,h)\}$ under the diffeomorhism $\chi_l \circ A_l$ in (4) which does not depend on h. We denote the kernel $K_n \circ \chi_\ell$ by $\tilde{K}_n$ and assume that x, the source point, is given by $x = \chi_\ell(u)$. Note that $\tilde{K}_n\psi$ is independent of h. As in the weakly singular case, we would like to devise one integration formula on the reference element Ω_0, by which all integrals of the form (29) can be computed accurately up to any required asymptotic order in h. Then all that is needed to obtain an integral over Ω_0 is to scale the variables in $\tilde{K}_n$ by h^{-1}, giving $\tilde{K}_n(hv_1, hv_2)$ on Ω_0. As is shown in [13, Theorem 2], in finite part integrals (29) in general there arise additional point functionals with changes of variables, i.e.

$$I_{jl}^h = \text{p.f.} \int_{\Omega_0^h} \tilde{K}_n(u; v)\psi(v)|\frac{\partial\chi}{\partial v}|dv + \sum_{0 \leq |\alpha| \leq n} c_\alpha \frac{1}{\alpha!}(D^\alpha u)(0). \tag{30}$$

However, as we explained in Theorem 1, for every source point x inside the patch F_{jl}^h the constants c_α are zero. For the approximation of

$$I^h = \text{p.f.} \int_{\Omega_0^h} \tilde{K}_n(u; v)\psi(v)|\frac{\partial\chi}{\partial v}|dv$$

we introduce polar coordinates $v_1 = r\cos\theta, v_2 = r\sin\theta$ in the parameter plane and we assume further that the shape functions on Γ are scaled with the Jacobian $|\frac{\partial\chi}{\partial v}|$, i.e. that

$$I_{jl}^h = \text{p.f.} \int_{F_{jl}^h} K_n(x, y - x)\psi(\chi_l^{-1}(y))|\frac{\partial\chi}{\partial v}|^{-1}ds_y.$$

Let us now explain which quadrature formulas we shall use to approximate

$$I^h[\tilde{K}_{-2-n}\psi] = \text{p.f.} \int_0^\omega \int_0^{hR(\theta)} (\tilde{K}_n\psi)(u; r\cos\theta, r\sin\theta)r\,dr\,d\theta.$$

In $\theta-$direction we shall use a Gauss-Formula G^M, properly scaled:

$$G^M[\tilde{K}_{-2-n}\psi] = \sum_{i=1}^{M} w_i^M \text{ p.f.} \int_0^{hR(\theta_i)} (\tilde{K}_n\psi)(u; r\cos\theta_i, r\sin\theta_i) r\, dr.$$

To compute the inner $r-$integral, we shall use an interpolatory finite part formula of degree N which is different for each θ_i. It is determined as follows:

i) For $\theta = \theta_i$ fix $0 < r_1 < \cdots < r_N \leq R(\theta_i)$.

ii) Find $w_j(\theta)$ such that for

$$Q_\theta^N[P] = \sum_1^N w_j(\theta) P(r_j)$$

there holds:

$$Q_\theta^N[P] = \text{p.f.}_{\varepsilon\to 0} \int_\varepsilon^{R(\theta)} r^{-1-n} P(r) dr \qquad \text{for all } P \in \Pi_{N-1}(0, R(\theta))$$

where $\prod_{N-1}$ denotes the set of all polynomials of degree $N-1$.

The determination of the weights $w_j(\theta)$ amounts to the solution of a system of N linear equations. Thus,it is necessary to insure that the combined formula $G^M Q^N$ approximates the finite part on Ω_0 obtained by removing an $\varepsilon-$circle about 0.
We point out that, since the source points are usually images of the same point $u \in \Omega_0$, the equations for determining the weights need only be solved once and the weights may be provided a-priori in practical implementation.

Let us now present an error estimate for the combined formula $G^M Q^N$ applied to the integrand $\tilde{K}_n(u; v-u)\psi$, i.e. the kernel with the parameter representation χ of Γ.

Theorem 7:
Let the boundary parametrization $(R(\theta), \theta)$ of $\partial\Omega_0$ be analytic on $[0,\omega]$. Then there holds

$$\begin{aligned} &|I^h - G^M Q^N[(\tilde{K}_n u)(hv_1, hv_2)] - \sum_{0\leq|\alpha|\leq n} c_\alpha(h)\frac{1}{\alpha!}(D^\alpha u)(0)| \\ &\leq C_N h^{-n}\left\{(1+\delta_{on}|\ln h|)e^{-2M\ln\delta} + h^N\right\}. \end{aligned}$$

Here I^h denotes the finite part integral (30), $c_\alpha(h)$ are the point functionals arising from scaling the domain Ω_0^h by $\frac{1}{h}$, δ_{ij} is the Kronecker symbol, and δ is a constant independent of h.

REFERENCES

[1] ALIABADI, M. H., HALL, W. S.: "Weighted Gaussian methods for there dimensional boundary element kernel integration". Comm.Appl.Num.Math., 30 (1987) pp. 89–96.

[2] COSTABEL, M., WENDLAND, W. L.: "Strong ellipticity of boundary integral operators". Journ. Reine Angew. Math., 372 (1986) pp. 34–63

[3] DUFFY, M. G.: "Quadrature over a pyramid or cube of integrals with a singularity at a vertex", SIAM J. Numer.Anal., 6 (1982) pp. 1260–1262.

[4] GRAY, L. J., MARTHA, L. F., INGRAFFEA, A. R.: "Hypersingular integrals in boundary element fracture analysis", Intern. J. Numer. Methods Engrg., 29 (1990) pp. 1135–1158.

[5] GUIGGIANI, M., GIGANTE, A.: "A general algorithm for multidimensional Cauchy principal value integrals in the boundary element method", ASME J.Applied Mechanics, 57 (1990) pp. 907–915.

[6] GUIGGIANI, M., KRISHNASAMY, G., RUDOLPHI, T. J., RIZZO, F. J.: "A general algarithm for numerical solulion of hypersingular boundary integral equations", ASME J.Applied Mechanics, to appear.

[7] KIESER, R.: „Über einseitige Sprungrelationen und hypersinguläre Operatoren in der Methode der Randelemente“, Doctoral Dissertation, University Stuttgart 1991.

[8] LYNESS, J. N.: "An error functional expansion for N-dimensional quadrature with an integrand function singular at a point", Math. Comp., 30 (1976) pp. 1–23.

[9] LYNESS, J. N.: "Applications of extrapolation techniques to multidimensional quadrature of some integrand functions with a singularity", Journ. Comp. Phys., 20 (1976) pp. 346–364.

[10] LYNESS, J. N.: "Quadrature error functional expansions for the simplex when the integrand function has singularities at vertices", Math. Comp., 34 (1980) pp. 213–225.

[11] NEČAS, J.: "Les méthodes directes en théorie des équations elliptiques". 1st Ed., Masson, Paris, 1967.

[12] RABINOWITZ, P., RICHTER, N.: "New error coefficients for estimating quadrature errors for analytic functions", Math. Comp., 24 (1970) pp. 561–570

[13] SCHWAB, C., WENDLAND, W. L.: "Kernel properties and representation of boundary integral operators", to appear. (Preprint Univ.Stuttgart 1990.)

[14] SCHWAB, C., WENDLAND, W. L.: "On numerical cubatures of singular surface integrals in boundary element methods", to appear. (Preprint Univ.Stuttgart 1990.)

[15] STROUD, A. H.: "Approximate Calculation of Multiple Integrals", Prentice Hall, Englewood Cliffs NJ, 1971.

[16] WENDLAND, W. L.: "Strongly elliptic boundary integral equations", In: The State of the Art in Numerical Analysis (A. Iserles and M. Powell eds.), Clarendon Press, Oxford (1987), pp. 511–561.

[17] WENDLAND, W. L.: „Analysis und Numerik von Randelementmethoden“, ZAMM, to appear.

List of Participants:

Andrä, Dr. H. Institut für Technische Mechanik, Festigkeitslehre, Universität Karlsruhe, Postfach 6980, D-W-7500 Karlsruhe 1

Antes, Prof. Dr. H. TU Braunschweig, Inst. für Angew. Mechanik, Mendelssohnstr. 3, D-W-3300 Braunschweig

Burmeister, Jens Institut für Informatik und Praktische Mathematik, Universität Kiel, Olshausenstr. 40, D-W-2300 Kiel 1

Casciola, C.M. INSEAN, Via di Vallerano 139, I-00184 Roma, Italy

Dallner, Rudolf Lehrstuhl für Technische Mechanik, Universität Erlangen-Nürnberg, Am Pestalozziring 20, D-W-8520 Erlangen

Dehnhardt, Jürgen Angewandte Mathematik, Universität Hannover, Welfengarten 1, D-W-3000 Hannover 1

Engl, Dipl.-Math. Gabriele Mathematisches Institut, Technische Universität München, Arcisstr. 21, Postfach 202420, D-W-8000 München

Gaul, Dr.-Ing. L. Universität der Bundeswehr Hamburg, Institut für Mechanik, Postfach 70 08 22, Holstenhofweg 85, D-W-2000 Hamburg 70

Georgiev, Dr. K. Center for Informatics and Computer Technology, Acad. G. Bonchevstr. , bl. 25-A, BG-1113 Sofia, Bulgaria

Graf, Kai JAFO-Technologie, Postfach 80 05 27, D-W-2000 Hamburg 80

Guiggiani, M. Dipartimento di Costruzioni, Meccaniche e Nucleari, Universita degli Studi di Pisa, via Diotisalvi 2, I-56126 Pisa, Italy

Haack, C. Technische Universität, Hamburg-Harburg, Arbeitsbereich Meerestechnik II, Eißendorfer Str. 42, D-W-2100 Hamburg 90

Hackbusch, Prof. Dr. W. Institut für Informatik und Praktische Mathematik, Universität Kiel, Olshausenstr. 40, D-W-2300 Kiel 1

Hartmann, Prof. Dr. F. Wipfelweg 60, D-W-4600 Dortmund 30

Holzer, S. Institut für Bauingenieurwesen I, Technische Universität München, Arcisstr. 21, D-W-8000 München 2

Jäger, Jürgen Steuben Str. 50, D-W-6900 Heidelberg

Jäger, Monika IBM Deutschland, Wissenschaftliches Zentrum Heidelberg, Tiergartenstr. 15, D-W-6900 Heidelberg

Jensen, Gerhard Hamburgische Schiffbauversuchsanstalt, Bramfelder Str. 164, D-W-2000 Hamburg 60

Kalik, Dr. Karl Universität Stuttgart, Mathematisches Institut A, Pfaffenwaldring 57, D-W-7000 Stuttgart 80

Kapust, Birgit Institut für Informatik und Praktische Mathematik, Universität Kiel, Olshausenstr. 40-60, D-W-2300 Kiel 1

Katzer, Dr. Edgar Institut für Informatik und Praktische Mathematik, Universität Kiel, Olshausenstr. 40, D-W-2300 Kiel 1

Kieser, Ralf Mathematisches Institut A, Universität Stuttgart, Pfaffenwaldring 57, D-W-7000 Stuttgart 80

Knöpke, Bernd GH Kassel, FB 14, Mönckebergstr. 7, D-W-3500 Kassel

Kolodziej, Dr. hab. Jan A. Technical University of Poznan, Institut of Applied Mechanics, Piotrowo 3, PL-60-965 Poznan, Polen

Krätzschmar, Prof. Dr. Michael Institut für Angewandte Mathematik, Fachhochschule Flensburg, Kanzleistr. 91-93, D-W-2390 Flensburg

Lancia, Dr. M.R. Dip. Meccanicaa e Aeronautica, Via Goossiania 18, I-00184 Roma, Italy

Leinen, Dr. Peter Mathematisches Institut, Auf der Morgenstelle 10, Universität Tübingen, D-W-7400 Tübingen

Liebau, Frank Institut für Informatik und Praktische Mathematik, Universität Kiel, Olshausenstr. 40, D-W-2300 Kiel 1

Lubich, Dr. Ch. Inst. f. Mathematik, Universität Innsbruck, Technikerstr. 13, A - 6020 Innsbruck

Lyness, Prof. James N. Mathematics and Computer Science Division, Argonne National Laboratory, 9700 South Cass Avenue, Argonne IL 60439, U.S.A.

Mahrenholtz, Prof. Dr.-Ing. Oskar Arbeitsbereich Meerestechnik II -, Strukturmechanik, TU Hamburg-Harburg, Eißendorfer Str. 42, D-W-2010 Hamburg 90

Matusiak, Jerzy Senior Research Scientist, Technical Research Centre of Finland, (VTT) Ship Laboratory PL 112, SF-02151 Espoo, Finland

Pomp Universität Stuttgart, D-W-7000 Stuttgart

Prößdorf, Prof. Dr. S. Karl-Weierstraß-Institut für Mathematik, Mohrenstr. 39, D-O-1086 Berlin

Pylkkänen, Jaakko VTT Ship Laboratory, P.O.Box 112, SF-02151 ESPOO, Finland

Rieder, Dr. Georg Sandweg 37, D-W-5100 Aachen-Laurensberg

Röttgermann, Andreas Universität der Bundeswehr, Werner-Heisenberg-Weg 39, D-W-8014 Neubiberg

Sauter, Stefan Institut für Informatik und Praktische Mathematik, Universität Kiel, Olshausenstr. 40, D-W-2300 Kiel 1

Schanz, M. Institut für Mechanik, Universität der Bundeswehr Hamburg, Holstenhofweg 85, D-W-2000 Hamburg 70

Schieweck, Dr. F. Sektion Mathematik, TU 'Otto von Guericke' Magdeburg, PSF 124, D-O-3010 Magdeburg

Schippers, Dr. H. National Aerospace Laboratory NLR, Anthony Fokkerweg 2, NL-8300 Amsterdam Emmeloord

Schlegel, Dr. V. Techn. Universität Hamburg-Harburg, Arbeitsbereich Meerestechnik II, Eißendorfer Str. 42, D-W-2100 Hamburg 90

Schlüter, Dr. Hans-Joachim Institut für Informatik und Praktische Mathematik, Universität Kiel, Olshausenstr. 40, D-W-2300 Kiel 1

Schmidt, Thorsten Institut für Informatik und Praktische Mathematik, Universität Kiel, Olshausenstr. 40, D-W-2300 Kiel 1

Schnack, Prof. Dr.-Ing. E. Institut f. Technische Mechanik, Universität Karlsruhe, Kaiserstr. 12, D-W-7500 Karlsruhe 1

Schneider, Dr. Reinhold FB Mathematik, THD-Darmstadt, Schloßgartenstr. 7, D-W-6100 Darmstadt

Stephan, Prof. Dr. Ernst P. Angewandte Mathematik, Universität Hannover, Welfengarten 1, D-W-3000 Hannover 1

Streckwall, Heinrich Hamburgische Schiffbau-Versuchsanstalt GmbH, Bramfelder Str. 164, D-W-2000 Hamburg 60

Strese, Dr. H. Institut für Mathematik, Mohrenstr. 39, D-O-1086 Berlin

Thiede, Dr.-Ing. Rolf Curt-Risch-Institut für Dynamik, Schall- und Meßtechnik der Universität Hannover, Appelstr. 9 A, D-W-3000 Hannover 1

Tobiska, Prof. Dr. L. TU 'Otto von Guericke' Magdeburg, Sektion Mathematik, PSF 124, D-O-3010 Magdeburg

Volk, Dipl.-Ing. Andrea Hofrat-Steiner-Weg 2, D-W-6114 Groß-Umstadt

Volk, Dr. Klaus IBM Deutschland, Wissenschaftliches Zentrum Heidelberg, Tiergartenstr. 15, D-W-6900 Heidelberg

Wendland, Prof. Dr. W. Mathematisches Institut A, Universität Stuttgart, Pfaffenwaldring 57, D-W-7000 Stuttgart 80

Wittum, Dr. Gabriel SFB 123 Universität Heidelberg, Im Neuenheimer Feld 294, D-W-6900 Heidelberg

Xiao, Dr. Lin Mechanik, RWTH Aachen, Templergraben 64, D-W-5100 Aachen

Addresses of the Editors of the Series "Notes on Numerical Fluid Mechanics"

Prof. Dr. Ernst Heinrich Hirschel (General Editor)
Herzog-Heinrich-Weg 6
D-8011 Zorneding
Federal Republic of Germany

Prof. Dr. Kozo Fujii
High-Speed Aerodynamics Div.
The ISAS
Yoshinodai 3-1-1, Sagamihara
Kanagawa 229
Japan

Prof. Dr. Bram van Leer
Department of Aerospace Engineering
The University of Michigan
Ann Arbor, MI 48109-2140
USA

Prof. Dr. Keith William Morton
Oxford University Computing Laboratory
Numerical Analysis Group
8-11 Keble Road
Oxford OX1 3QD
Great Britain

Prof. Dr. Maurizio Pandolfi
Dipartimento di Ingegneria Aeronautica e Spaziale
Politecnico di Torino
Corso Duca Degli Abruzzi, 24
I-10129 Torino
Italy

Prof. Dr. Arthur Rizzi
FFA Stockholm
Box 11021
S-16111 Bromma 11
Sweden

Dr. Bernard Roux
Institut de Mécanique des Fluides
Laboratoire Associé au C. R. N. S. LA 03
1, Rue Honnorat
F-13003 Marseille
France

Brief Instruction for Authors

Manuscripts should have well over 100 pages. As they will be reproduced photomechanically they should be typed with utmost care on special stationary which will be supplied on request.
In print, the size will be reduced linearly to approximately 75 per cent. Figures and diagrams should be lettered accordingly so as to produce letters not smaller than 2 mm in print. The same is valid for handwritten formulae. Manuscripts (in English) or proposals should be sent to the general editor, Prof. Dr. E. H. Hirschel, Herzog-Heinrich-Weg 6, D-8011 Zorneding.